Renewable Energies for Central Asia Countries: Economic, Environmental and Social Impacts

NATO Science Series

A Series presenting the results of scientific meetings supported under the NATO Science Programme.

The Series is published by IOS Press, Amsterdam, and Springer (formerly Kluwer Academic Publishers) in conjunction with the NATO Public Diplomacy Division.

Sub-Series

I. Life and Behavioural Sciences	IOS Press
II. Mathematics, Physics and Chemistry	Springer (formerly Kluwer Academic Publishers)
III. Computer and Systems Science	IOS Press
IV. Earth and Environmental Sciences	Springer (formerly Kluwer Academic Publishers)

The NATO Science Series continues the series of books published formerly as the NATO ASI Series.

The NATO Science Programme offers support for collaboration in civil science between scientists of countries of the Euro-Atlantic Partnership Council. The types of scientific meeting generally supported are "Advanced Study Institutes" and "Advanced Research Workshops", and the NATO Science Series collects together the results of these meetings. The meetings are co-organized by scientists from NATO countries and scientists from NATO's Partner countries — countries of the CIS and Central and Eastern Europe.

Advanced Study Institutes are high-level tutorial courses offering in-depth study of latest advances in a field.
Advanced Research Workshops are expert meetings aimed at critical assessment of a field, and identification of directions for future action.

As a consequence of the restructuring of the NATO Science Programme in 1999, the NATO Science Series was re-organized to the four sub-series noted above. Please consult the following web sites for information on previous volumes published in the Series.

http://www.nato.int/science
http://www.springeronline.com
http://www.iospress.nl

Series IV: Earth and Environmental Sciences – Vol. 59

Renewable Energies for Central Asia Countries: Economic, Environmental and Social Impacts

edited by

Aldo Iacomelli
Pisa University - ISES ITALIA - Italian Section of International Solar Energy Society, Rome, Italy

Springer

Proceedings of the NATO SFP Workshop on
Renewable Energies for Central Asia Countries: Economic, Environmental and Social

A C.I.P. Catalogue record for this book is available from the Library of Congress.

ISBN 1-4020-3924-7 (HB)
ISBN 978-1-4020-3924-9 (HB)
ISBN 1-4020-3926-3 (e-book)
ISBN 978-1-4020-3926-3 (e-book)

Published by Springer,
P.O. Box 17, 3300 AA Dordrecht, The Netherlands.

www.springeronline.com

Printed on acid-free paper

Printed in the Netherlands.

TABLE OF CONTENTS

ACKNOWLEDGEMENTS

I would like to thank the NATO Scientific Division, in particular Dr. Alain Jubier, in supporting the initiative; the Uzbekistan Authorities, for the tireless cooperation, the help and the kindness in hosting the workshop, in particular Professor Alik Akunov, Dr. Svetlana Gusakova, the academician Pulat Kabibullaeve.

I would like to thank dr. Giacomo Spaghetti for the translation from Russian during the workshop and for the support in the organization. Many thanks to the authors of this book and to dr. Marzia Tamburrino and Mrs Paola Carducci for their useful help in editing this book, and dr. Emanuele Piccinno and dr. Daniele Villoresi for the suggestions and the help in the administration part of the project.

I would like to thank ISES ITALY staff and Professor Vincenzo Naso, CIRPS Director, who made possible the mission to Uzbekistan. Finally I would like to thank Dr. Alicia Mignone, Italian OCSE representative, for the important help in identify, with dr. Carlo Corsi, director of Consorzio Roma Ricerche, the donor, NATO Scientific Division, and for the help to design the workshop program.

Last but not least I would like to thanks Leonardo and Marta that are waiting for me continuously and patiently for the meetings and the books.

TASHKENT STATEMENT

**of the NATO Advanced Research Workshop
«Advanced Renewable Energies for Central Asia Countries:
economic, environmental and social impacts»**

On November 15-17, 2004 the workshop was held in Tashkent organized by the NATO Science Committee in association with the Center for Science and Technologies under the Cabinet of Ministers of the Republic of Uzbekistan, the Inter-University Research Center on Sustainable Development (CIRPS), University of Rome "LA SAPIENZA" (Italy) and Technology Transfer Agency (Uzbekistan).

The workshop was attended by representatives of NATO Science Committee, scientists, specialists and experts from the USA, UK, Canada, France, Italy, Japan, Australia, Russia, Ukraine, Kazakhstan, Kyrgyzstan and Uzbekistan and United Nations Environment Programme Risoe Centre - Denmark.

The analysis, study and evaluation of all aspects of the energy sector in Uzbekistan and other Central Asian countries, especially from the point of development prospects and utilization of the renewable energy sources, their potentials in positive influence on economic development were regarded as important goal of cooperation between NATO member countries and Central Asian countries.

The participants of the workshop recognized that:

The International NATO Science Committee workshop was organized together with the Center for Science and Technologies under the Cabinet of Ministers of the Republic of Uzbekistan and Inter-University Research Center on Sustainable Development (CIRPS) and University of Rome "LA SAPIENZA" (Italy) and Technology Transfer Agency.

Uzbekistan has significant scientific, technical and technological potential in the field of advanced renewable energy technologies including a unique 1 MW Big Solar Furnace for high temperature synthesis of materials with given properties as well as for generating electric, thermal energy and hydrogen. Scientific and technical achievements in developing solar elements based on semi-conductor crystals, experience gained in using low-grade solar energy, results of high temperature materials engineering. These achievements can contribute to a partnership between NATO countries and Republic of Uzbekistan and other Central Asian countries.

The achievements of the workshop were:

mutual benefits, scientific and practical effectiveness significantly contributing to innovation processes of the Republic of Uzbekistan. The workshop encouraged establishment and further development of the mutually beneficial contacts between scientific representatives of NATO member and partner countries.

The participants of the workshop recommended that:

1. The Center for Science and Technologies under the Cabinet of Ministers of the Republic of Uzbekistan and Technology Transfer Agency to approve of a cooperation program with NATO Science Committee, including grants for implementing innovation research achievements in developing technologies in the field of renewable energy, as well as solar energy, in Uzbekistan and Central Asian countries.

2. The Center for Science and Technologies under the Cabinet of Ministers of the Republic of Uzbekistan to organize and conduct a series of workshop in the regions of Uzbekistan in order to disseminate experience gained in the field of advanced renewable energy technologies and stimulating innovation activity.

3. a program of collaboration between the Republic of Uzbekistan, Central Asian and Newly Independent countries and NATO member countries in information and expertise exchange related to implement institutional and legal framework for using renewable energy sources could be an important condition to develop innovation processes.

4. The Inter-University Research Center on Sustainable Development CIRPS, University of Rome "La Sapienza" recommends to Central Asian countries to work out collaborative and prospective research and technological projects on renewable energy with provisions for financing respective part from each country and to request assistance in

additional funding such projects from NATO Science Committee in form of grants. Organize next workshop on hydrogen energy in Tashkent in 2005, where would be discussed different methods of hydrogen production and storage; transport issues; evaluation of production costs; experience and operating data from demonstration projects; international and regional initiatives; national research programs and industry involvement.

The workshop could be organized by the Center for Science and Technologies under the Cabinet of Ministers of the Republic of Uzbekistan in collaboration with

5. The enter into force of Kyoto Protocol by spring 2005, after the ratification of more than 55% of emissions from Annex 1 countries, with the Protocol mechanisms such as Joint Implementation (JI), Clean Development Mechanism (CDM) and Emission Trading (ET) will offer an important opportunity for Governments and industries to use RES and technological innovation in the energy sector to meet Kyoto target in the first budget period, to improve quality of life, economic growth and to reduce emissions.

6. To ensure that our research activity, which is our collective strength, will be used for constructive partnership for change and for the achievement of the common goal of promoting the rational use of energy, energy saving and the promotion of Renewable Energy towards a sustainable development;

7. To support the initiative by working with the financial community to facilitate an increase in lending to, and investment in, renewable projects using the proposed mechanisms. Parallel activities will also be undertaken to help financial institutions become more aware of RE investment opportunities, streamline procedures, lower transaction costs, and assess/manage the risks and returns associated with the deployment of the selected technologies.

8. CIRPS, University of Rome "La Sapienza" to lead the creation of a network of scientists, representatives of administrative structure and private sector to stimulate research activities in the renewable energies, energy efficiency and hydrogen economy and technologies.

9. It is recommended to publish proceedings of the NATO ARW «Advanced technologies «Advanced Renewable Energies for

Central Asia Countries: economic, environmental and social impacts» in a separate book.

Signatures

On behalf of
the Republic of Uzbekistan

Academician Pulat Khabibullaev
Director of Center
for Science and Technologies
under the Cabinet of Ministers
of the Republic of Uzbekistan

Professor Aldo Iacomelli
General Secretary
of International Solar Energy Society
Italian Section

Academician Vincenzo Naso
CIRPS Director General
University of Rome "La Sapienza"

Tashkent, Republic of Uzbekistan 17 November 2004

LIST OF AUTHORS

Aldo Iacomelli
Pisa University
Secretary General of ISES ITALIA Co-director NATO Advanced Research Workshop
Professor
Via E. Filiberto, 1-3
56127 Pisa
Mob. +39 335 8141630
Tel. +39 050 2213363
e-mail: aldo.iacomelli@uniroma1.it

Via Tomasso Grossi, 6
00184 Rome
Mob. +39 335 8141630
Fax +39 06 77073612
Tel. +39 06 77073610 –11
e-mail: sg@isesitalia.it

PACUDAN Romeo
United Nations Environmental Programme RISO Center
Doctor
P.O. Box 49 DK- 4000 Roskilde, Denmark
Phone: +45 4677 5170 Fax: +45 4632 1999
E-mail: romeo.pacudan@risoe.dk

MIGNONE Alicia
Energy and Science Advisor Permanent Delegation of Italy to OECD
Professor
50, Rue de Varenne 75007 Paris, France
Tel: (33) 01 44 39 21 60 Fax: (33) 01 45 48 00 60
Email: alicia.mignone@rappocse.org

MALYSHEV Teresa Carroll
Renewable Energy export IEA International Energy Agency,
Energy Analyst
9, rue de la Federation 75739 PARIS Cedex 15
Tel: (33) 1 40 57 67 12 Fax: (33) 1 40 57 66 59
E-mail: teresa.malyshev@iea.org

BARONI Marco
International Energy Agency
Doctor
9, rue de la Federation 75739 PARIS Cedex 15
Tel: +33 1 40 57 65 82 Fax: +33 1 40 57 66 59
E-mail: marco.baroni@iea.org

KOCH Frans Herman
Executive Committee Implementing Agreement for Hydropower Technologies and Programmes
Doctor
26 Meadowcroft Cr., Ottawa, Ont K1J1G9,Canada
Tel.: +(1) 613 744 56 11 Fax: +(1) 613 748 31 57
E-mail: fkoch@gvsc.ca

NASO Vincenzo
CIRPS Director General University of Rome "La Sapienza"
Academician
CIRPS – Piazza del Colosseo, 9 -00184 Roma – Italia
Tel.: +39 06 772653206 Fax: +39 06 772653215
Email: cirpsest@uniroma1.it

NASYROV Timur Khayrullaevich
Director of Energy Center of Uzbekistan
Academician
Tel/fax: +998 71 133 08 15, E-mail: uzem@sarkor.uz

MOLODTSOV Sergei Dmitrievich
Deputy Director for Science of the Center for Energy Policy of Russia
Tel. 8 095 200 45 06, 200 37 34, 200 44 79
E-mail: aslan@energy.ru

ABDULLAEV Djura Abdullaevich
Scientific advisor to Center of scientific-engineering and marketing research of Uzbek Agency of communication and information
Academician
Tel + 99871 137 56 24, Fax +99871 137 52 07,
E-mail: ftmtm@uzpak.uz

ISAEV Rikhsi Isakhodjaevich
Deputy Director of Center of Scientific Engineering and Marketing Research of the Uzbek Agency
of Communication and Information
Tel (998 71) 137 56 24, Fax (998 71) 137 52 07,
E-mail: ftmtm@uzpak.uz

MUMINOV Ramizulla Abdullaevich
Physics and Technical Institute of NPO "Physics-Sun"
Academician
Tel +998 71 135 40 32 Fax +998 71 135 42 91

NOSOVA Ludmila
Thermal Physics Department of Uzbek Academy of Sciences
Tel. +998 71 117 48 70 Fax: +998 71 117 48 71

SAIDOV Abdulla Abdunazarovich
Deputy Head of Heat Physics Department of the Academy of Sciences of the Republic of Uzbekistan
Professor
Tel. +998 71 117 48 70, 173 37 96 Fax: +998 71 117 48 71

KHABIBULLAEV Pulat Kirgizbaevich
Director of Center for Science and Technologies (CST) under the Cabinet of Ministers of the Republic of Uzbekistan

Director of Thermal Physics Department of Uzbek Academy of Sciences
Academician
Fax +998 71 137 61 78

KOYSHIEV Temirkhan Kosybaevich
Head of Renewable Energy Sources Department of Kazakh State University
Professor
Almaty, Si-Sinhay str.22, apt.48,
tel 8 3272 48 27 65 mobile 8 333 23 150 75
E-mail: galim-galim@mail.ru, temirluch@ok.kz

KACHKYNBAEVA Kulsina Kanietovna
Director of Branch of Central Asia Regional Environment Center in Kyrgyz Republic
Bishkek, Aliaskar Toktonalieva str 2/1, Kyrgyzsatn. 720055
tel./fax. (8 10 996 312) 61 13 55, E-mail: rec-ulsina@mecd.bishkek.gov.kg, KKachkynbaeva@carec.kz, carec-kg@mail.kg

LEBEDEV Oleg Vladimirovich
Head of Laboratory of Institute of Construction Mechanics and Seismic Resistance of Academy of Science of Uzbekistan
Chief of Department of Tashkent Automotive Institute
Academician
E-mail: leb_oleg@yahoo.com.

LIST OF PARTICIPANTS

NATO COUNTRIES

C A N A D A

KOCH Frans Herman
Executive Committee Implementing Agreement for Hydropower Technologies and Programmes
Doctor
26 Meadowcroft Cr., Ottawa, Ont K1J1G9,Canada
Tel.: +(1) 613 744 56 11 Fax: +(1) 613 748 31 57 E-mail: fkoch@gvsc.ca

D E N M A R K

PACUDAN Romeo
United Nations Environmental Programme RISO Center
Doctor
P.O. Box 49 DK- 4000 Roskilde, Denmark
Phone: +45 4677 5170 Fax: +45 4632 1999
E-mail: romeo.pacudan@risoe.dk

F R A N C E

BARONI Marco
International Energy Agency
Doctor
9, rue de la Federation 75739 PARIS Cedex 15
Tel: +33 1 40 57 65 82 Fax: +33 1 40 57 66 59
E-mail: marco.baroni@iea.org

I T A L Y

IACOMELLI Aldo
Inter-University Consortium For Research on Sustainable Development (CIRPS)
University of Rome "La Sapienza"
Secretary General of ISES ITALIA Co-director NATO Advanced Research Workshop
Professor
Via E. Filiberto, 1-3
56127 Pisa
Mob. +39 335 8141630
Tel. +39 050 2213363
e-mail: aldo.iacomelli@uniroma1.it

Via Tomasso Grossi, 6
00184 Rome
Mob. +39 335 8141630
Fax +39 06 77073612
Tel. +39 06 77073610 – 11
e-mail: sg@isesitalia.it

NASO Vincenzo
CIRPS Director General University of Rome “La Sapienza”
Academician
CIRPS – Piazza del Colosseo, 9 -00184 Roma – Italia
Tel.: +39 06 772653206 Fax: +39 06 772653215
Email: cirpsest@uniroma1.it

SPAGHETTI Giacomo
CIRPS - University of Rome “La Sapienza”
Doctor
CIRPS – Piazza del Colosseo, 9 -00184 Roma – Italia
Tel.: +39 06 772653206 Fax: +39 06 772653215 Mob.: 3493423348
Email: cirpsest@uniroma1.it

MIGNONE Alicia
Energy and Science Advisor Permanent Delegation of Italy to OECD
Professor
50, Rue de Varenne 75007 Paris, France
Tel: (33) 01 44 39 21 60 Fax: (33) 01 45 48 00 60
Email: Alicia.mignone@rappocse.org

SWEDEN

TELENIUS Bjorn
Executive Committee Implementing Agreement for Bioenergy, International Energy Agency
National Energy Administration
Biomass expert
Box 310, 631 04 Eskilstuna Besoksadress: Kungsgatan 43
Tel: 016-544 2109 Fax: 016-544 2261 Mobil: 070-497 0401
E-mail: bjorn.telenius@stem.se

UNITED STATES

MALYSHEV Teresa Carroll
Renewable Energy export IEA International Energy Agency,
Energy Analyst
9, rue de la Federation 75739 PARIS Cedex 15
Tel: (33) 1 40 57 67 12 Fax: (33) 1 40 57 66 59
E-mail: teresa.malyshev@iea.org

NIS COUNTRIES

KAZAKHSTAN

KOYSHIEV Temirkhan Kosybaevich
Head of Renewable Energy Sources Department of Kazakh State University
Professor
Almaty, Si-Sinhay str.22, apt.48,
tel 8 3272 48 27 65 mobile 8 333 23 150 75
E-mail: galim-galim@mail.ru, temirluch@ok.kz

KYRGYZSTAN

KACHKYNBAEVA Kulsina Kanietovna
Director of Branch of Central Asia Regional Environment Center in Kyrgyz Republic
Bishkek, Aliaskar Toktonalieva str 2/1, Kyrgyzsatn. 720055
tel./fax. (8 10 996 312) 61 13 55, E-mail: rec-ulsina@mecd.bishkek.gov.kg, KKachkynbaeva@carec.kz, carec-kg@mail.kg

OBOZOV Alaybek Jumabekovich
Director of "Kyun" Renewable Energy Center under the Government of Kyrgyz Republic
Member of Engineering Academy
Professor
Tel. (8 10 996 312) 55 92 01 / 05, Fax 8 10 996 312 55 92 04,
Mob. 0502 57 02 66
E-mail: kun@elcat.kg

RUSSIA

MOLODTSOV Sergei Dmitrievich
Deputy Director for Science of the Center for Energy Policy of Russia
Tel. 8 095 200 45 06, 200 37 34, 200 44 79
E-mail: aslan@energy.ru

STREBKOV Dmitry Semenovich
Director of All-Union Institute of Agriculture Electrification
Academician
Tel. 8 095 170 51 01, 171 19 20 E-mail: viesh@dol.ru

KHARCHENO Valery Vladimirovich
Director of All-Union Institute of Agriculture Electrification
Head of RES dept
Doctor

Tel. (+7095) 171 1423, (+7095) 171 2191, Fax. (+7095) 171 5101
E-mail: viesh@dol.ru; kharval@mail.ru

UKRAINE

KUDRYA Stepan Aleksandrovich
Deputy Director for science of the Renewable Energy Institute of the National Academy of Sciences of Ukraine
Professor
Tel/fax 8 10 38 (044) 559 23 63 Mobile 8 10 38 (067) 465 66 68
E-mail: mntc@mbox.com.ua

UZBEKISTAN

PARTICIPANTS

KHABIBULLAEV Pulat Kirgizbaevich
Director of Center for Science and Technologies (CST) under the Cabinet of Ministers of the Republic of Uzbekistan
Director of Thermal Physics Department of Uzbek Academy of Sciences
Academician
Fax +998 71 137 61 78

AKHUNOV Ali Akhunovich
Deputy Director of Center for Science and Technologies under the Cabinet of Ministers of the Republic of Uzbekistan
Professor
Tel. +998 71 135 62 70 Fax +99871 137 62 74

ABDULLAEV Djura Abdullaevich
Scientific advisor to Center of scientific-engineering and marketing research of Uzbek Agency of communication and information
Academician
Tel + 99871 137 56 24, Fax +99871 137 52 07,
E-mail: ftmtm@uzpak.uz

MUMINOV Ramizulla Abdullaevich
Physics and Technical Institute of NPO "Physics-Sun"
Academician
Tel +998 71 135 40 32 Fax +998 71 135 42 91

NASYROV Timur Khayrullaevich
Director of Energy Center of Uzbekistan
Academician

Tel/fax: +998 71 133 08 15, E-mail: uzem@sarkor.uz

ABDURAKHMANOV Abdujabor Abdurakhmanovich
Director of Institute of Materials of NPO "Physics- Sun"
Professor
Tel/fax +998 71 133 95 64 Tel +998 272 224 10

ALLAEV Kahramon Rakhimovich
Rector of Tashkent State Technical University
Professor
Tel. +998 71 144 78 80. Fax +998 71 118 90 52

ZAKHIDOV Romen Abdullaevich
Head of Laboratory of renewable energy sources of the Institute of Energy and Automatics of the Uzbek Academy of Sciences.
Head of theoretical heat technics of Tashkent State Technical University
Academician
Tel: +998 71 162 09 21, E-mail: zakhidov@uzsci.net

LEBEDEV Oleg Vladimirovich
Head of Laboratory of Institute of Construction Mechanics and Seismic Resistance of Academy of Science of Uzbekistan
Chief of Department of Tashkent Automotive Institute
Academician
E-mail: leb_oleg@yahoo.com.

LUTPULLAEV Sagdulla Lutpullaevich
Director of Physics-Technical Institute NPO "Physics- Sun"
Professor
Tel. +998 71 133 12 71 Fax +998 71 135 42 91

SAIDOV Abdulla Abdunazarovich
Deputy Head of Heat Physics Department of the Academy of Sciences of the Republic of Uzbekistan
Professor
Tel. +998 71 117 48 70, 173 37 96 Fax: +998 71 117 48 71

ISAEV Rikhsi Isakhodjaevich
Deputy Director of Center of Scientific Engineering and Marketing Research of the Uzbek Agency
of Communication and Information
Tel (998 71) 137 56 24, Fax (998 71) 137 52 07,

E-mail: ftmtm@uzpak.uz

GUSAKOVA Svetlana Dmitrievna
Director of Technology Transfer Agency
Co-Director of NATO ARW
Doctor
Tel/Fax +998 71 139 49 17, Tel 39 48 07
E-mail: s.gusakova2004@mail.ru

NOSOVA Ludmila
Thermal Physics Department of Uzbek Academy of Sciences
Tel. +998 71 117 48 70 Fax: +998 71 117 48 71

INVITED PERSONS

SHIRIN Vadim Viktorovich
Director of Qurilishgelioservice Corporation
Tel. . +99871 135 41 60, +99871133 64 47,
E-mail qgs@uzsci.net,

NIKULINA Svetlana Petrovna
Programme Manager of Environment Programme of the Government of Uzbekistan supported by UNDP
e-mail svtelana.nikulina@envp.uzsci.net

AZAROV Oleg Vladimirovich
Head of energy saving technologies department of National Holding Company “Uzneftegaz”
Tel. +99871 136 37 43

INTRODUCTION

In the last few years, the awareness and worries towards the exhaustion of natural resources are increasing, but this fact is still very discussed about its analytical measurement, and therefore is not yet demonstrated.

However, what has been demonstrated is that the 6 billion inhabitants of the Earth move around the world every kind of goods and wares for a level of 8 tons per-capita, which means about 50 billion tons per year. This is the same amount as that of the materials moved yearly by natural forces, such as winds, eruptions, earthquakes, rain.

The Man, with his anthropic activities, became a "geologic force". The most important environmental problems caused by human activities, are connected with the use of natural resources.

The United Nations, with the IPCC (International Panel on Climate Change), studied the climate changes, and demonstrated, in their scientific reports, that:

- In the last 100 years of the 1900, the Earth temperature rose, in average, of 0.4 - 0.6 °C;
- The XX century was the warmest of the last six centuries, and the last years of 1900 were the warmest;
- By the end of the next century, is expected an increase of the temperature, in average of 2 °C, with greater increase at higher latitudes.
- The level of the sea has risen of an average of about 10-25 cm, in particular after the thermo-expansion of the oceans. The temperature of the sea is rising also at the deepest levels.

Because of all this facts, an intervention on greenhouse gas emissions is more and more pressing: even with an immediate intervention, the temperature of the Earth would increase for decades, due to the entrapment of solar heat in the atmosphere.

Today the concentration of carbon dioxide is of more than 358 ppm (part per million), and is 30% higher compared to 200 years ago (pre-industrial era).

The dependence from fossil sources is still too strong, and the mix of combustibles in the future will be more and more dependent from the OPEC countries, in particular those in the Middle East, as well as from some NIS countries in the Caspian region.

The shift to coal tries to soften this geopolitical dependence from fossil sources, but does not respond properly to climate crisis and to health and environmental emergencies. Consumption of fossil sources are divided mainly into three macro-sectors: production of thermo-electric energy, mobility of goods and people, heating/conditioning and lighting of rooms.

The world average of thermo-electric power plants are very old, and there were no investments for many years, today the average efficiency of electricity production is about 35 – 37 %. New power plants with combined cycle, reach an average efficiency of 55%, but with the distributed cogeneration, small power plants near by energy demand it would be possible to obtain even higher performances with less costs included the environmental costs.

In the residential field, there is an average consumption of about 200 kWh/m^2/year, in the offices and private houses compared to 50 kWh/m^2/year in some countries where there is an investment in energy demand side management, with systems of passive heating, day-lighting, could be used less than one third of that amount of energy word wide.

However, the mobility sector (mainly road transport) of goods and people, is the sector that is mostly dependent on fossil combustible, oil and its by-products, with massive rigidity of the system and with an increase of the rate of growth of consumptions and gas emissions of about 27% yearly world wide.

An investment on rational uses of energy, savings and efficiency is the main premise to support the development of new energy sources its needed. If energy consumption decreases, renewable sources could cover a significant part of the demand of energy (in particular electricity), if consumption remains uselessly high because inefficient, renewable energy sources reduce their positive impact.

If many phases of industrial production (even heavy industries) become more efficient and less energy-consuming (acting also on final uses), renewable energy would become a reality, a feasible method even in these sectors.

With investments being equal (today all in the sector of generation from fossil sources), if there were parallel researches on how to reduce consumption and wastes considerably (at least 35%) and on power plants from renewable sources, there would be also a reduction of gas emissions, without any negative influence on development.

New technologies (and new “energy products”) will play a crucial role for the development of a market of “sustainable energy products” that should grow in a competitive way (cost-effective) to stand against the challenge of change.

Aldo Iacomelli, PhD
Editor

Chapter 1

RENEWABLE ENERGY (RE), ENERGY EFFICIENCY (EE) & ENERGY SERVICES: THE ENERGY MARKET TRANSFORMATION

Aldo Iacomelli, PhD

Pisa University -
ISES ITALIA – Italian Section of International Solar Energy Society
iacomelli@scamb.unipi.it

1. THE "PLAN OF IMPLEMENTATION" OF THE JOHANNESBURG WORLD SUMMIT ON SUSTAINABLE DEVELOPMENT (WSSD)

The WSSD in Johannesburg strongly reaffirms the commitment to the Rio principles, the full implementation of Agenda 21 and the Programme for the Further Implementation of Agenda 21, and also commit the nations to achieving the internationally agreed development goals, including those contained in the United Nations Millennium Declaration trough the "Plan of Implementation". To this end, concrete actions and measures are needed at all levels and to enhancing international cooperation, taking into account **the principle of common but differentiated responsibilities** as set out in **Principle 7 of the Rio Declaration on Environment and Development.**

The three components of sustainable development as interdependent and mutually reinforcing pillars, as confirmed outcome of WSSD are:

- ***economic development;***
- ***social development;***
- ***environmental protection.***

A. Iacomelli (ed.), Renewable Energies for Central Asia Countries: Economic Environmental and Social Impacts, 1–17.

After Johannesburg the "*new paradigma*", showed in the diagram below, keeps jointed the three pillars with the social issues on the top of Sustainable Development Pyramid as the new driving force for the development, and definitely clarifies the huge difference between the economic growth (the previous paradigma in the business as usual age) and the new perspectives of durable development: the challenge for the future generations.

Fig 1. Johannesburg vs. Rio the Janeiro the Pyramid of Priority

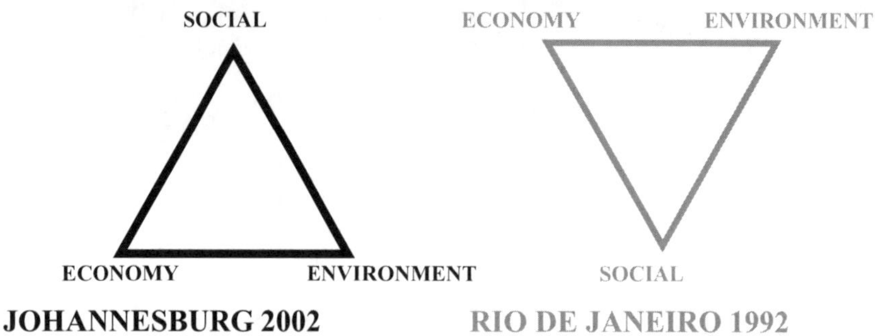

Poverty eradication, changing unsustainable patterns of production and consumption, and protecting and managing the natural resource base of economic and social development are overarching objectives of, and essential requirements for, sustainable development.

At the international and national level for achieving sustainable development and ensuring SOCIAL, ECONOMIC and ENVIRONMENTAL, essential benefits are:

- ***peace;***
- ***security (food, water, energy, information health services, culture);***
- ***stability and respect for human rights and fundamental freedoms;***
- ***social and economic policies;***
- ***democratic institutions responsive to the needs of the people;***
- ***respect for cultural diversity.***
- ***the right to development;***
- ***gender equality;***
- ***the rule of law, anti-corruption measures;***
- ***good governance within each country;***

♦ ***sound environmental practices.***

we will focus, to the end of this article on energy issues and we will loop to the WSSD outcome on Energy and Transport

1.1 Energy

The energy issues where the weak point in Johannesburg Summit. The nations weren't able to achieve an agreement with target and deadline for needed actions to prevent climate changes and to avoid energy shocks. Nevertheless one of the outcome is to increase the global share of renewable energy sources with the objective of increasing its contribution to total energy supply, recognizing the role of national and voluntary regional targets as well as initiatives. Develop and utilize indigenous energy sources and infrastructures for various local uses and promote rural community participation, in developing and utilizing renewable energy technologies to find simple and local solutions. To facilitate the access to energy as one of the measures to poverty eradication actions at all levels should include:

- ➢ *improve access to reliable, affordable, economically viable, socially acceptable and environmentally sound energy services and resources,*
- ➢ *enhanced rural electrification and decentralized energy systems,*
- ➢ *increased use of renewables,*
- ➢ *cleaner liquid and gaseous fuels*
- ➢ *enhanced energy efficiency,*
- ➢ *capacity-building,*
- ➢ *financial and innovative financing mechanisms;*
- ➢ *technological assistance;*
- ➢ *improve access to modern biomass technologies and fuel-wood sources and supplies;*
- ➢ *support the transition to the cleaner use of liquid and gaseous fossil fuels, where considered more environmentally sound, socially acceptable and cost-effective;*
- ➢ *develop national energy policies and regulatory frameworks;*
- ➢ *create the necessary economic, social and institutional conditions in the energy sector to improve access to reliable, affordable, economically viable, socially acceptable and environmentally sound energy services;*
- ➢ *promote public-private partnerships.*

In Johannesburg the focus was not only on renewable energy technologies but also on taking into account cleaner fossil fuel technologies and, under pressure of many developing countries, large hydro, with the aim of a transfer to developing countries on concessional terms as mutually agreed. But if we combine the WSSD outcome with the Kyoto Protocol target, because it is a legal binding instrument, RE&EE and sustainable energy are the future of development and not the clean fossil fuel technology.

Promote increased research and development in the field of new energy service, including renewable energy, energy efficiency and advanced sustainable energy technologies, based specifically on gas CHP and microcogen. *Diversify energy supply by developing advanced, cleaner, more efficient, affordable and cost-effective energy technologies.*

The technology transfer must be realized on a mutual agreed bases between developed and developing countries. This actions should include:

- develop domestic programmes for energy efficiency;
- provide environmentally sound technology transfer;
- eliminate the barrier for the diffusion of environmentally sound technologies;
- develop capacity-building at national and regional level;
- develop and disseminate alternative energy technologies with the aim of giving a greater share of the energy mix to renewable energies, energy technologies, including cleaner fossil fuel technologies as transition frap;
- promote BAT and BEP in the energy sector;
- promote research programme for further development in energy technology;
- provide technical assistance to developing countries, with the involvement of the private sector, to reduce flaring and venting of gas associated with crude oil production.

1.2 Transport

Promote an integrated approach to policy-making at the national, regional and local levels for transport services and systems to promote sustainable development, including policies and planning for land use, infrastructure, public transport systems and goods delivery networks, with a view to providing:

1. *safe, affordable and efficient transportation;*
2. *increasing energy efficiency;*
3. *reducing pollution;*
4. *reducing congestion;*
5. *reducing adverse health effects;*
6. *limiting urban sprawl.*

This would include actions at all levels to:

- ➢ implement transport strategies for sustainable development, improve the affordability, efficiency and convenience of transportation;
- ➢ improve urban air quality and health,
- ➢ reduce greenhouse gas emissions, development of better vehicle technologies, more environmentally sound, affordable and socially acceptable;
- ➢ promote investment and partnerships for the development of sustainable, energy efficient multi-modal transportation systems;
- ➢ promote public mass transportation systems;
- ➢ promote better transportation systems in rural areas;
- ➢ ensure technical and financial assistance for developing countries and countries with economies in transition.

1.3 Public-Private Partnership (PPP)

Some concrete elements for the development of means of implementation work program, to make concrete the outcome of WSSD.

Promote public-private partnerships, financial support for technical assistance, the development of technology and capacity-building to developing countries to:

- ❖ enhance trade infrastructure;
- ❖ strengthen financial institutions;
- ❖ increase developing country capacity to diversify and increase exports to cope with the instability of commodity prices and declining terms of trade;
- ❖ increase the value added of developing country exports.

Continue to enhance the mutual supportiveness of trade, environment and development with a view to achieving sustainable development through actions at all levels to:

- ✓ support the completion of the work programme of the Doha Ministerial Declaration on subsidies so as to promote sustainable development and enhance the environment;
- ✓ encourage reform of subsidies that have considerable negative effects on the environment and are incompatible with sustainable development;
- ✓ encourage efforts to promote cooperation on trade, environment and development,
- ✓ providing technical assistance to developing countries, between the secretariats of WTO, UNCTAD, UNDP, UNEP and other relevant international environmental and development and regional organizations;
- ✓ encourage the voluntary use of environmental impact assessments as an important national-level tool to better identify trade, environment and development inter-linkages.

1.4 Economic and financial measures

Provide financial assistance to developing countries, with the involvement of the private sector in the energy field with the aim of mobilizing the adequate financial instruments and resources. Develop and support efforts to improve transparency and information about energy markets.

The international community should develop policies and measures to reduce market distortions and promote energy systems compatible with sustainable development through the use of improved market signals and by removing market distortions, including restructuring taxation and phasing out harmful subsidies, where they exist. To this end actions is needed, where appropriate, to phase out subsidies in this area that inhibit sustainable development, taking fully into account the specific conditions and different levels of development of individual countries and considering their adverse effect, particularly on developing countries;

Recommend that international financial institutions and other agencies' policies support developing countries, as well as countries with economies in transition, in their own efforts to establish policy and regulatory frameworks which create a level playing field between:

1. *renewable energy,*
2. *energy efficiency,*
3. *advanced energy technologies (including advanced and cleaner fossil fuel technologies, and centralized, distributed and decentralized energy systems).*

Financial instruments and mechanisms, should utilize, also, the Global Environment Facility (GEF), within its mandate, to provide financial resources to developing countries, in particular least developed countries and small island developing States, to meet their capacity needs for:

- develop technical know-how;
- capacity building;
- training;
- strengthening national institutions in reliable, affordable, economically viable, socially acceptable and environmentally sound energy;
- promoting energy efficiency and conservation;
- promoting renewable energy;
- promoting advanced energy technologies (including advanced and cleaner fossil fuel technologies).

International and regional cooperation arrangements should be strengthened for promoting cross-border energy trade, including the interconnection of electricity grids and oil and natural gas pipelines.

2. STOP THE GROWTH OF GREENHOUSE GAS EMISSIONS WITH THE RATIONAL USE OF ENERGY AND OF NEW ENERGY SERVICES &SOURCES.

The governments should use several instruments to choose and support some energy technologies or innovative energy products (included savings and efficiency), such as:

- Fiscal measures;
- Policies and guidelines;
- Information, labeling, voluntary agreements, other programs of assistance and support;
- Research and developments of new technologies.

In particular, it is about technology progress, with the ecological modernization. These measures of "energy policy" should focus on the "amount of energy produced by renewable sources" to maximize it, and on "costs of production of energy from renewable sources" to minimize it.

Using price instruments, a "strategy" is used based on renewable energy sources, trying to create a relevant supply. For this purpose, a gross remuneration is assured to the producers, with a price guaranteed higher than that of energy from conventional sources, so that the differential revenue is an incentive to a greater opening of the market, from the supply side, and to the removal of every entry barrier.

After fixing the unit price, after necessary fluctuations, the quantity produced at that price will be defined freely and stabilized with time.

With the instruments of quantity, on the other hand, it is the amount of energy produced that is fixed, leaving the price floating to set at that level of production.

This happens with a mechanism of "competitive auction" that selects producers according to the lower costs.

The difference between the 2 options is the difference between price and costs. It is necessary to distinguish between short term and long term.

In the short term, the auction mechanism will lead to lower production costs for energy from renewable sources, because it removes the revenues, and therefore is to be considered more efficient. The beneficiaries of the missing burden of the revenue are the energy buyers, against who would have weight that burden, trough the dispenser utilities, or trough the general fiscal system.

In the short term as well, the quantity mechanism will lead, vice versa, to an extra price for the community, an extra price that should not be considered different from a public investment in a private research: the expressed aim is to recover that investment achieving a growing technical efficiency and, therefore, minor costs in the future.

This reflection passes on to the long term, when the possible re-investment of the differential revenues, caused by the mechanism of the guaranteed price, in technology research with results which can be useful to all producers, will lead gradually to lower costs.

Fig. 2. Technology learning curve & cost reduction opportunities

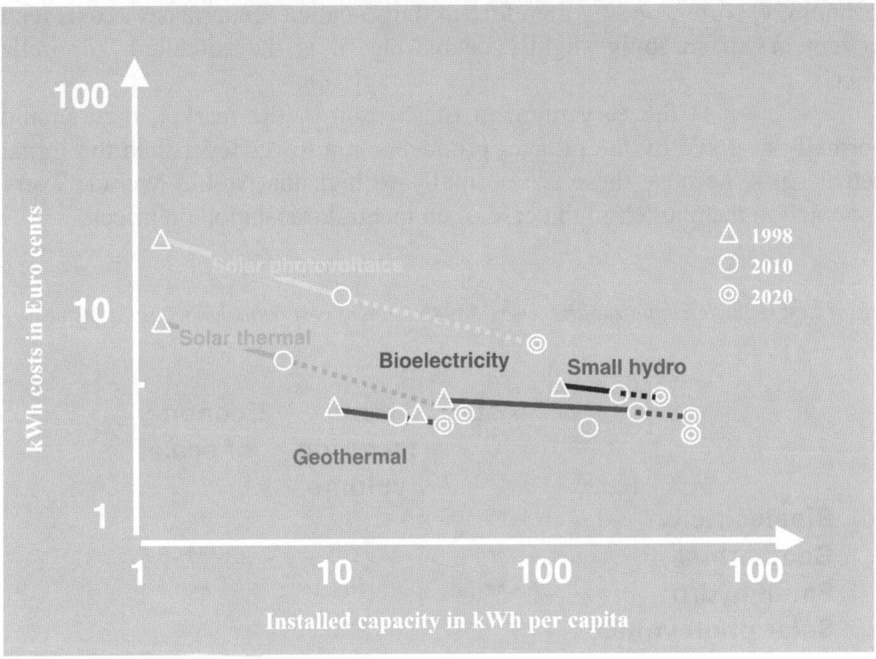

Source: elaboration from IEA data

Fig. 3. RE market costs-price evolution in the global context

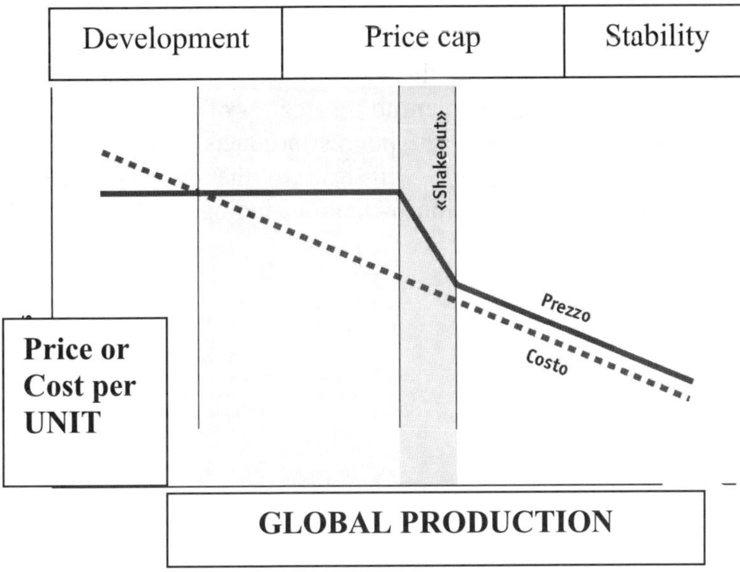

At the beginning of any period characterized by a certain technology, corresponding therefore to the so-called "first phase", costs tend to remain stiff or only slightly decreasing: it is the so-called "umbrella price".

At the very moment of creation of the market, they should normally be fixed by the pioneer producers at a lower lever than the initial actual costs, because these are normally so high that would exclude every approach to the market, in the case of an integral translation on prices.

Fig. 4 Cost reduction opportunities, every * = 4% - 6% of cost reduction within a decade.

	R&D	Manu-facturing volume	Economy of scale
Bioelectricity	**	*	*
Geothermal	**	*	***
Small hydro	**	*	**
Solar photovoltaics	*****	****	*
CSP	***	***	****
Wind onshore	**	*	***
Wind offshore	***	*	***

Source: elaboration from IEA data

For example, in the case of energy efficiency, the market mechanism entailed a displacement to a category of products with a greater efficiency and is marginalizing the poorest products. The transformation of the market of energy products will lead to the development of better technologies and deeper diffusion of new technologies, if compared to the "starting point".

Fig. 5 the role of "learning by doing" comes out from the efficiency lesson

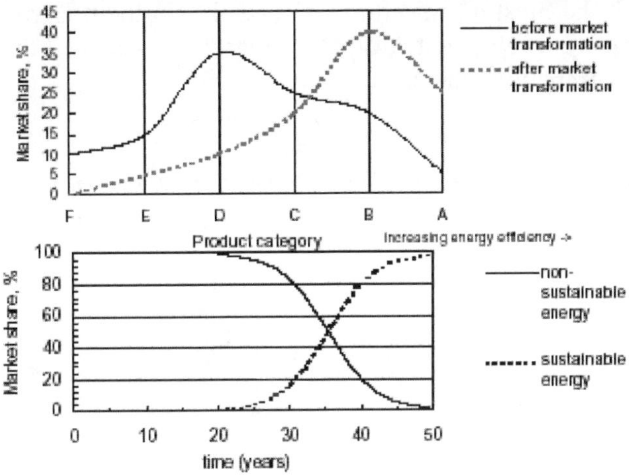

Fig. 6 Annual Growth of Renewables Supply from 1971 to 2000

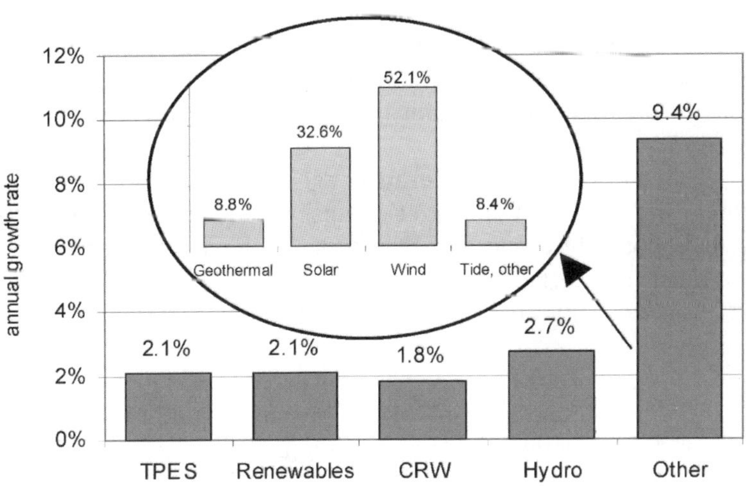

Innovative business strategies could facilitate the transformation of the energy market.

In a very simple way, until not long ago, the core business of an utility was the energy selling to domestic clients or to industries, and the key factor to success was the price of electricity. In fact, within a "regulated" market, any transformation of the market itself was practically not possible. The liberalization of electricity and gas is increasing the competitivity of the market. Looking at the most advanced experiences in Northern Europe (for example UK), the price of electricity at first decreased, almost as if the price itself was the only competitive factor.

Fig. 7 Technology push and market pull the pillar of the energy policy reform

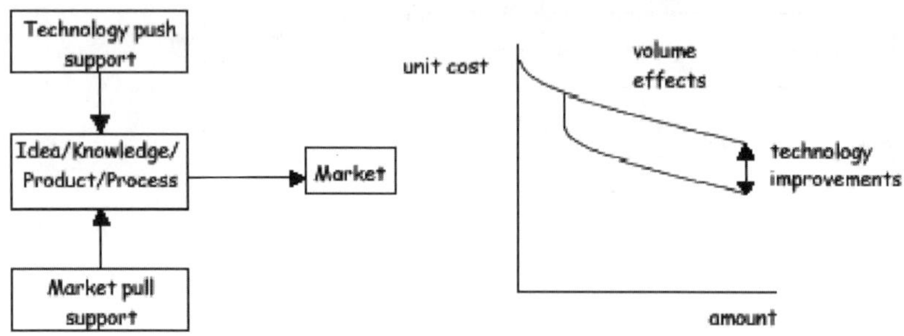

To the last extent, those companies with most competitive prices and the lowest costs will prevail in the market. But, without electricity at low cost, how can an energy company survive in the market, in the long term?

The Business is not only to sell a product or a technology but is something that involves the "target groups" of clients and its distribution channels. It is not only a question of new technologies, but a new way of offering products and services.

The utilities can have benefits on the market if they are able to be innovative.

Where do the money go in the cycle of energy "value"?

The "desegregation" of the business of utilities has started. Previously, when production and transmission (and also distribution) were in the same hands, the emphasis (even for the weight of prices) was on production. After the dismantling, still on the way, of the monopoly of the utilities in Europe, the "peak of price" is today within the line of distribution, and the final users are often "linked" to local distributors.

Move the attention, as is going to happen in the next future, on final users in a disaggregated market, much of the profits would come from services to final users.

This is going to fully happen when consumers could have the possibility to choose between options of distribution.

The road of the "innovative energy products" has been tracked: after a

The transition phase will be dominated by gas and carbon, than there are the renewable energies and especially the rational uses of energy will be a forced road. But, since the transition from an energy system to another will last decades, the foundation of a future system is built now. Therefore, a full application of the European set of laws needs also a strong political shove and investments to direct the technology innovation, the transformation of the energy market and the choices towards rational uses of energies and renewables.

This view is enriched also from the employment possibilities that, in the case of renewable energy, are an important added value.

Table 1. Workers from different source to produce annually a TWh of energy

Source	Workers (workers-year) (TWh)
Oil	260
Oil Off-Shore	265
Natural Gas	250
Coal	370
Nuclear	75
Wood	1000
Hydro	250
Mini – hydro	120
Wind	918
PV	76.000
Ethanol (from biomass)	4.000

Source: ISES ITALIA elaboration from lecterature

In addition to the important normative framework set by the European Union and by Italian institutions, we need to point out some instruments of energy policies that are able to stimulate the rational uses of energy and renewable sources, such as recommended in the conclusion.

3. CONCLUSION

The Policy & Measure suggested to develop in a country such as the central Asia Countries a path for renewable energy and energy efficiency should follow a set of guide principle such as:

- *ECONOMIC AND FINANCIAL ASPECTS*
  a) Incentives to production, as the "feed in tariffs or law", fiscal credits to production, "net metering", fiscal credits to the utilities that "sell" energy efficiency to final users;
  b) Financial mechanisms, as "bond", loans and mortgages at a lower rate, fiscal credits and mechanisms of public and private financial support to the production of energy from renewable sources.
  c) "benefit" systems (System Benefits Charges - SBC), to support the recover of financial or fiscal incentives, to cover within the shortest time the loans demanded by the entrepreneur of renewables energy.
  d) Programs of financing from Public bodies to University and specific agencies for the basic researches, and intervention of private capitals for applied researches and commercial development;
  e) Incentive to commercial mechanisms ("Trading") of "commodities", as green certificates, white certificates and Certificates of Emissions Reduction (CERs) fixed by the Kyoto Protocol, to strengthen the renewable energies, enhance the entry of this energy sources in the market, increase the value of the environmental added benefits of such energy choices.
  f) Introduction of target taxation (carbon energy tax) not only as an increased fiscal withdraw, but also as governmental "zero incentives" to the business and utilities that do not use renewable energy sources.
- *INTERVENTIONS TROUGH INTERMEDIATE AND LOCAL BODIES*
  g) individuation of minimum shares of renewable energies that every municipality, Province or Region in Italy should provide, requesting each local body to cover at least 35% of its energy needs with "green electricity", produced from renewable sources.
  h) Development of regional, provincial and municipal programs, to promote renewable sources and research applied to the land, together with Italian and European universities and companies.
- *INTERVENTION ON PROCEDURES AND AUTHORIZATION*
  i) Removal of procedural and bureaucratic difficulties, as building licenses and other authorizations that are a barrier for the use of renewable sources, always considering the need for the protection of

the environment and of the Italian artistic heritage, and especially the time limit for the decisions about authorizations.

j) Removal of extreme economic and fiscal barriers that hinder renewable sources, promoting access to the network for auto-producers, including individual users.

k) Standardization process of regulatory and authorization systems on a national basis, that would allow unifying them and with the national and European sets of laws

l) Development and application of agreements of inter-connection to the transmission network that have to be standardized and easy to apply.

REFERENCES

S. Awerbuch "Regulation and Pricing for Distribution: Promoting DG Through Cost-Based Access and Usage Charges," Public Utilities Fortnightly, July 1, 2000.

Beer, J. de, M.T. Van Wees, E. Worrell and K. Blok, 1994, *ICARUS-3: The potential for energy efficiency improvement in the Netherlands up to 2000 and 2015*, Department of Science, Technology and Society, Utrecht University, Utrecht.

Beer, J. de, E. Worrell and K. Blok, forthcoming, 'Future Technologies for Energy-Efficient Iron and Steel Making', accepted for publication in Annual Review of Energy and Environ-ment (23) 1998.

Bertoldi, P., 1996, *European Union Efforts to Promote More Efficient Use of Electricity: the PACE Programme*, European Commission, Brussels.

Bertoldi, P., 1996b, EC/DG-XII, Unit Electricity, Personal communication, November 12, 1996.

Bertoldi, P., 1998, EC/DG-XII, Unit Electricity, Personal communication, May 7, 1998.

Cogen, 1997, *European Cogeneration Review 1997*, Cogen Europe, Brussels, Belgium

EC, 1996a, *Energy for the Future: renewable sources of energy. Green Paper for a Community Strategy*, Communication from the Commission COM(96)576, Brussels, 1996.

EC, 1996b, *European Energy to 2020 - A Scenario Approach*, European Commission, Luxembourg, 1996.

EC, 1997b, *Proposal for a Council Directive restructuring the Community framework for the taxation of energy products*, European Commission, Brussels.

EC, 1997c, *A Community strategy to promote combined heat and power (CHP) and to dismantle barriers to its implementation,* Communication from the Commission to the Council and the European Parliament, COM(97)514 final, 15/10/97, Brussels.

EC, 1997d, *Energy for the future: Renewable sources of energy*, White Paper for a Community Strategy and Action Plan, European Commission, Brussels.

EC, 1998, *Communication on Transport and CO_2 - Developing a Community Approach,* European Commission, Brussels.

EC, 1998a, *Environmental Agreement with the European Automobile Manufacturers Association (ACEA) on the Reduction of CO_2 Emissions from Passenger Cars,* Commission Staff Working Paper, SEC(1998)1047, Commission of the European Communities, Brussels.

European Parliament, Session Document, "Report on the Proposal for a European Parliament and Council Directive in the promotion of electricity from renewable energy sources COM(2000) 279–C5–0281/2000–20000/0116(COD).

ESD, 1996, *TERES-II, The European Renewable Energy Study,* (prepared by Energy for Sustainable Development), European Commission, DGXVII (Altener Programme), Brussels.

Farla, J., K. Blok, L.J. Schipper, 1997, *Energy Efficiency Developments in the Pulp and Paper Industry - A Cross-Country Comparison Using Physical Production Data,* Energy Policy **25,** pp. 745-758.

FCCC. 1997, *Kyoto Protocol*, COP-3, UNFCCC, Kyoto.

GEA, 1995, *Washing machines, driers and dishwashers, final report*, Group on Efficient Appliances, publ. by Danish Energy Agency, Kopenhagen.

Gusbin, D., 1995; *Policies and Measures for "Common Action", Case study on product energy efficiency standards*, Coherence, Louvain-la-Neuve, Belgium.

Hendriks, C.A., J-W Velthijsen, E. Worrell, and K. Blok, 1995, *Regulation and Energy Conservation: the case of Combined Heat and Power in the European Union – Situation and Prospects, for Joule-II project "Benefits and Costs of Policy Instruments for Energy Conservation"*, Dept. of Science, Technology and Society, Utrecht University, The Netherlands

Holsteijn, van, and Kemna engineering consultants, 1996, *Sensitivity analysis of energy efficiency improvement for Washing Machines*, EC-DGXVII, Brussels.

Howarth, R.B. and L. Schipper, 1991, *Manufacturing energy use in eight OECD countries: trends through 1988*, Lawrence Berkeley National Laboratory, Berkeley.

IEA Energy Outlook 2002. International Energy Agency,Paris, 2002

March Council, 1997, *Council Conclusions*, EU Ministerial Environment Council, Brussels, March.

Ministry of Economic Affairs, 1995, *Long term agreements with the Dutch industrial sector on energy efficiency; initial results*, Dutch Ministry of Economic Affairs, The Hague.

Nørgard, J.S., 1989, "Low electric Appliances - Options for the future", in: T.B. Johansson, B. Bodlund, and R.H. Williams (eds.) *Electricity, Efficienct End-use and New Generation Technologies and their Planning Implications*, Lund University Press, Lund.

Peterson, W.S. and R.E. Miller, 1986, *Hall-Heroult Centennial: first century of aluminium process technology 1886-1986*, Pennsylvania.

Phylipsen, G.J.M., K. Blok, E. Worrell, 1998b, *Benchmarking the energy efficiency of the Dutch energy-intensive industry; a preliminary assessment of the effect on energy consumption and CO_2 emissions,* Department of Science, Technology and Society, Utrecht University.

Phylipsen, G.J.M., K. Blok, H. Merkus, 1997, *The Expert's Group work on EU Common and Coordinated Policies and Measures*, Dept. of Science, Technology and Society, Utrecht University, Utrecht.

Solomon Associates Ltd., 1995, *Wordwide Olefins Plant Performance Analysis, 1995*, Solo-mon Associates Ltd., Windsor.

UNDP World Energy Assessment 2000. United Nations Development Programme, Washington, 2000

Utrecht University, 1997, *Evaluatie Meerjarenafspraken over energie-efficiency* (Evaluation long-term agreements on energy efficiency, in Dutch), by M.C. Das, P.P.J. Driessen, P. Glasbergen, N. Habermehl, W.J.V. Vermeulen, K. Blok, J.C.M. Farla and E.M. Korevaar, vakgroep Milieukunde, vakgroep Natuurwetenschap en Samenleving, Utrecht University, Utrecht.

WEC, 1993, *Renewable energy sources: opportunities and constraints 1990-2020,* World Energy Council, London.

Worrell, E., L. Price, N. Martin, J. Farla, and R. Schaeffer, 1997, *Energy Intensity in the Iron and Steel Industry: a comparison of physical and economic indicators,* in the Special Issue on Cross-country Comparisons of Indicators of Energy Use, Energy Efficiency and CO_2 Emissions of *Energy Policy*, June/July, pp.727-744.

Eurostat, 1999, "Energy: Yearly statistics 1998", European Communities, Luxembourg.

Krause F.,1999, "La Risorsa Efficienza. Strategie ed interventi per la riduzione delle emissioni di gas ad effetto serra attraverso misure di efficienza negli usi finali di energia elettrica", ANPA, Serie Documenti 11/1999.

IPCC/OECD/IEA, 1997, "Revised 1996 IPCC Guidelines for National Greenhouse Gas Inventories".

ODYSSEE Database, 1999, "Database on European Energy Efficiency Indicators" (http:// www.enerdata.grenet.fr/odyssee).

Chapter 2

PROMOTING EFFECTIVE AND EFFICIENT PUBLIC PRIVATE PARTNERSHIPS (PPPs)

Aldo Iacomelli, PhD

Pisa University -
ISES ITALIA – Italian Section of
International Solar Energy Society
iacomelli@scamb.unipi.it

1. INTRODUCTION

How to implement on the ground concrete projects? Which capital to support the investment and the management? From the WSSD summit one of the proposal was to promote public-private partnerships (PPP), to identify new financial support for technology innovation, development of this new sustainable technology and implement the needed capacity-building to help developing countries to:

- ❖ enhance trade infrastructure;
- ❖ strengthen financial institutions;
- ❖ increase developing country capacity to diversify and increase exports to cope with the instability of commodity prices and declining terms of trade;
- ❖ increase the value added of developing country exports.

Continue to enhance the mutual supportiveness of trade, environment and development with a view to achieving sustainable development through actions at all levels to:

- ✓ support the completion of the work programme of the Doha Ministerial Declaration on subsidies so as to promote sustainable development and enhance the environment;

A. Iacomelli (ed.), Renewable Energies for Central Asia Countries: Economic, Environmental and Social Impacts, 19–26.

- ✓ encourage reform of subsidies that have considerable negative effects on the environment and are incompatible with sustainable development;
- ✓ encourage efforts to promote cooperation on trade, environment and development,
- ✓ providing technical assistance to developing countries, between the secretariats of WTO, UNCTAD, UNDP, UNEP and other relevant international environmental and development and regional organizations;
- ✓ encourage the voluntary use of environmental impact assessments as an important national-level tool to better identify trade, environment and development inter-linkages.

2. PROPOSED FINANCIAL MECHANISM

The proposed financial mechanism, including the creation of a dedicated public-private "energy sector investment fund" (equity funds, multilateral funds, etc) will:

- ❖ create a more favourable public-private financing mechanism that will directly and indirectly increase the capability of our resources' capability to leverage private and other public resources;
- ❖ create a pool of patient capital, by allowing returns on investment which are below what the market, i.e. 5% instead of 15-25 %, that venture capitalists are seeking;
- ❖ increasing access to affordable capital by a special agreement with banks using the EIAF research analysis and evaluation (both technical and financial);
- ❖ promote the long-term viability of energy sector investments and growth perspectives of Energy Service Companies (ESCo) and related private sector activities in developing countries;
- ❖ complement awareness raising and capacity building efforts;
- ❖ offering increased access to a knowledgeable team of investors that can also provide managerial support to for developing country entrepreneurs (trough joint ventures and EIAF assistance).

2.1 Investment dedicated funds

The creation of an intermediate dedicated fund-based financial mechanism, would significantly increase the effectiveness and efficiency of public resources supporting "energy sector" investments. A fund-based mechanism can help to avoid a proliferation of fund application conditions and long time to deliver funds.

Fund based investment mechanism provide a more stable and reliable source of funding compared to resources made available on an annual basis. Fund-based resources are also more visible and more easily identifiable by potentially interested investors.

Financing energy sector ventures capital, in particular in developing countries will often require tailor made solutions. Selected public and private initiators could established a legal instrument setting out the rules of procedure governing a trust fund. The instrument would define the conditions under which public and private donors and investors could pool financial resources and build-up a viable portfolio of diversified investments.

2.2 Engaging pro-active fund managers

Investment financing is often a highly interactive process whereby decisions mainly take place in a commercial environment.

For the purpose of sound financial management, the investment management function could be split between an "operational" function (to be attributed to a professional investment management team such as the "Mediterranean Investment Advisory Facility"(EIAF)) and the "treasury and control", which could be attributed to the treasury of a well-respected public financial institution such as the ADB, EIB, the World Bank or the GEF.

2.3 Enhancing risk management options

The high risks associated often with investments in developing countries and to some extent (although of decreasing importance) associated with energy & water sector investments is a major obstacle for creating an "investment pipeline" towards sustainable energy research implementation in developing countries.

The proposed use of a dedicated revolving fund offers some important risk mitigation options for public and private donors. EIAF should train in developing countries "ad hoc" staffed fund management team that can be able to follow up the research implementation.

3. EXAMPLES OF ENERGY SECTOR PPP

The potential to leverage private financial resources is related to the nature of the investment, with the governance structures that will be put in place. It could be envisaged that the low return accruing to public investors is leave to private investors thereby offering a higher return than what they could get themselves in similar conditions.

Assuming that the investment fund is targeting middle income developing countries where expected returns are around 6% and the minimum ROI a private sector investor is requiring is 15%, the ratio of public versus private funding of such a PPP investment fund would be 60% - 40%.

This ratio would inversely relate to the expected returns from the private sector.

Fig. 1. Options for a Public Private Partnership

Public Private Partnership / Patient Capital			
Public	1.000.000	1%	0,0%
Private	99.000.000	99%	15,0%
	100.000.000	100%	14,9%
Public	60.000.000	60%	0,0%
Private	40.000.000	40%	15,0%
	100.000.000	100%	6,0%
Public	80.000.000	80%	0,0%
Private	20.000.000	20%	15,0%
	100.000.000	100%	3,0%

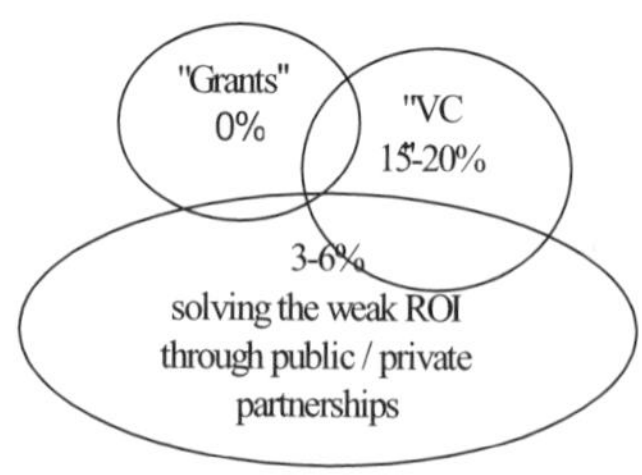

Depending on the assumptions taken, the leverage potential – measured in terms of the resources needed to generate a single "energy sector" investment capability, range from 10 to 160 in a non-leveraged fund structure.

3.1 Characteristics of "Project financing"

"An operation of project financing is the organization to finance a research, identifying the subjects that if there are appropriate guarantees, would entrust the repayment of their credit to the cash flows expected to be created by the research itself".

This type of operation has some characteristics:

- project financing itself is not a true financial instrument, but more a mixture of other instruments, already existing on the market;
- the receiver of the financing is the project itself, partly independent from the makers.

To complete a project financing there is the need to legally divide the research itself from the organizers;

- project financing is different from other financial instruments also because of the high number of bodies and organizations involved, each with diverse purposes and with specific risks.

To solve the problem of finding a financing with the instrument of "project financing", some conditions should be fulfilled, such as:

- there must be a project, defined by its performance objective, and by its time and resource limit;
- there must be some subjects that are determined to take individually the risks related to the realization of the project itself;
- these subjects must be able to structure a system of guarantees to support the financing of the project.

Without all these conditions, the use of a project financing is not possible. The main function of this kind of operation is the possibility to distribute properly the risks related to the project that, individually, none of the organizers would be able to tolerate.

These risks are of various nature:

- *technical risks, related to the making and management of the project;*
- *economical risks, related to the ineffective and inefficient use of the capital;*
- *operative risk, originating from the ineffective and inefficient use of the capital;*
- *market risks, caused by the demand fluctuation;*
- *institutional risks, that include also the so-called country risk, deriving from the low stability of political institutions in*

particular situations in the country (such as in many developing countries);

- *credit risk, caused by the possible insolvency of one or more promoters or guarantor;*
- *Financial risks, related to the fluctuation of interest, change and inflation rates.*

These risks are present in a different way in the different phases of the project, so that the financing bodies and the promoters of the project will face the risks depending on the phase in which they are more involved, based on the bonds they took.

In the end, it can be said that is possible to apply the project financing to the energy sector, as long as four conditions are fulfilled:

- *several institutions and organizations must have an interest in the project to be completed, because there is the need to share the risks related to the project;*
- *it must be possible to value in a correct way the dimension of the project that has to be financed, if the value of the energy infrastructures that are planned is not relevant compared to the debt capacity of the manager, this instrument is not appropriate, being sufficient more traditional financial instruments;*
- *complexity and technical characteristics of the project financing require significant management skills.*

A main issue is the division of the project into several phases, each financed by different investors with different risks.

The main variable in time, that require three phases:

- defining of the project and building of the works. It includes all engineering and planning works;
- starting of the project. It measures the suitability of the relation between the costs (also expected) and the specifications estimated during the planning of the financing;
- management as planned. Now the financial instrument acts as a real company. It is expected that cash flows coming from the service management can cover up operative costs, allowing at the same time a proper profit and the cover of the bonds originated by the contracts made to share the risks.

Fig. 2 Benchmark of successful "Project Financing" in the water sector (that conceptually is similar to the energy): "CONSORZIO AGUA AZUL SA"

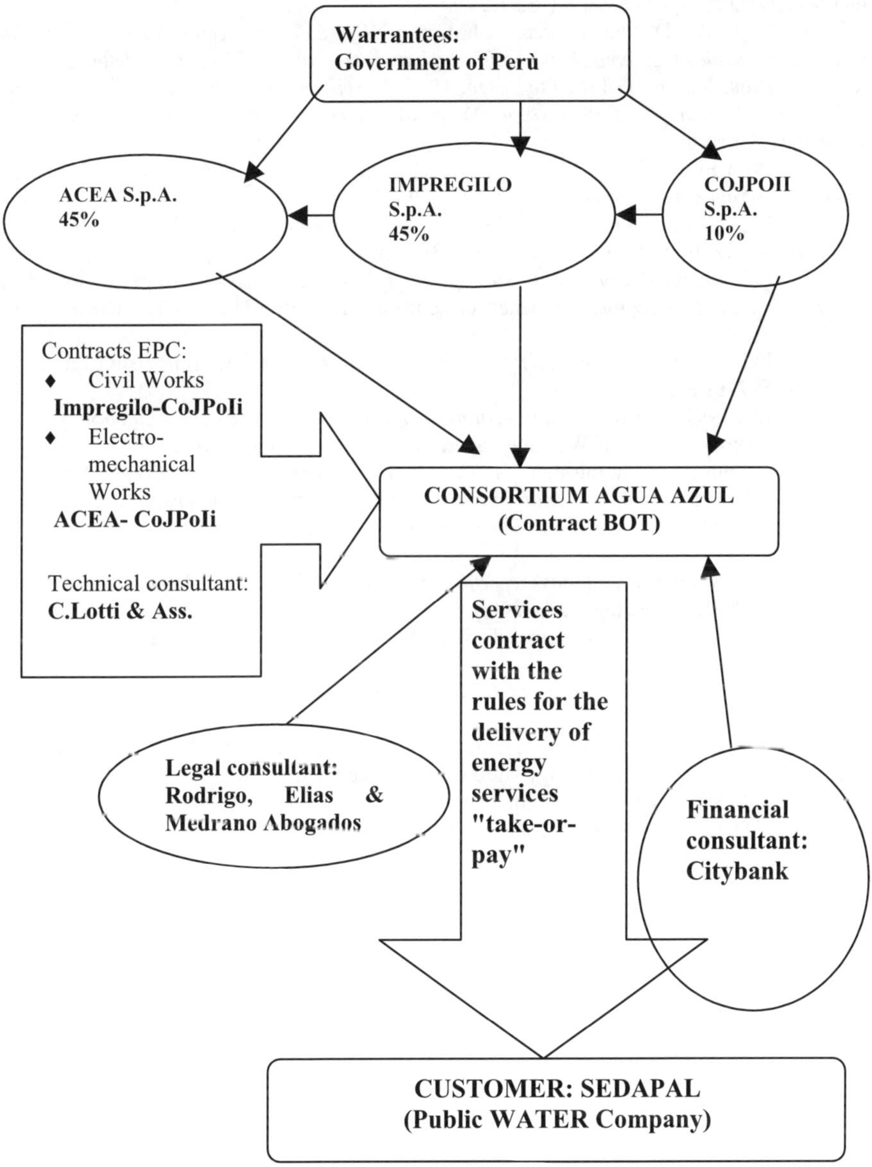

REFERENCES

Blok, K., and G.J.M. Phylipsen, 1996, *Common policies and measures for greenhouse gas emission limitation and reduction*, Background document for the workshop 'Towards a European consensus', Dublin, September 2-3, Department of Science, Technology and Society, Utrecht University, Utrecht.

Blok, K., D. van Vuuren, A.J.M. van Wijk and L.G. Hein, 1996, *Policies and measures to reduce CO_2 emissions by efficiency and renewables*, WWF Netherlands, Zeist.

Blok, K., and G.J.M. Phylipsen, 1997, *European Union policies and measures to achieve the 15% negotiating position*, Dept. of Science, technology and Society, Utrecht University, Utrrecht.

Burniaux, J.M., J.P. Martin, G. Nicoletti and J. Oliveira-Martins, 1992, *The costs of international agreements to reduce CO_2 emissions*, OECD, Paris.

Capros, P., T. Georgakopoulos, D. Van Regemoorter and S. Proost, 1996, Results from the General Equilibrium model GEM-E3, September.

EC, 1996c, *A Community strategy to reduce CO_2 emissions from passenger cars and improve the fuel economy*, Council Conclusions, European Union - The Council, 8748-96, Brussels.

EC, 1996d, *Strategy Paper for reducing methane emissions*, European Commission, COM(96)557, Brussels.

EC, 1997, *Climate change - Analysis of proposed EU emission reduction objectives for Kyoto*, Commission Staff Working Paper, European Commission, Brussels.

Enquete Kommission, 1995, *Mehr Zukünft für die Erde, Nachhaltige Energipolitik für dauerhafte Klimaschutz (More future for the earth; lasting energy policies for sustainable climate protection)*, Economica Verlag, Bonn.

IEA Energy and Poverty 2002. Special Publication for Johannesburg Summit International Energy Agency, Paris, 2002

IEA Energy Balances of non-OECD Countries 2001. International Energy Agency, 2002

IEA Energy Balances of OECD Countries 2001. International Energy Agency, 2002

Wijk, A.J.M. van, J.P. Coelingh, 1993, *Wind power potential in OECD countries,* Department of Science, Technology and Society, Utrecht University.

WSSD Plan of Implementation, advanced unedited text. World Summit on Sustainable Development, Johannesburg, 4 September 2002

Chapter 3

THE CLEAN DEVELOPMENT MECHANISM: NEW INSTRUMENT IN FINANCING RENEWABLE ENERGY TECHNOLOGIES

Romeo Pacudan, PhD

Senior Energy Economist
Risoe National Laboratory
Roskilde, Denmark

The Clean Development Mechanism (CDM) is one of the flexible mechanisms established under the Kyoto Protocol of the United Nations Framework Convention on Climate Change (UNFCCC) to assist industrialized countries in meeting their emissions reduction obligations at lower cost and at the same time to stimulate investments that promote sustainable development in developing countries. The UNFCCC is an international treaty formulated in 1992 and entered into force in 1994, which sets a goal of stabilizing atmospheric concentration of greenhouse gases at safe levels. The UNFCCC's supreme body, the Conference of Parties (COP), supervises the activities towards the achievement of the Convention's goals. In its first meeting in Berlin, Germany, the body decided that the post-2000 commitments to reduce emissions would only be set for industrialized countries, also known as Annex 1 countries. During the body's third meeting in Kyoto, Japan, the supreme body set a legally binding requirement for Annex-1 countries to trim down their greenhouse gas emissions to an average of 5.2% below their 1990 emissions levels. This legally binding commitment is also known as the Kyoto Protocol. In order for the Kyoto Protocol to enter into force, it requires ratification of at least 55 parties to the convention, which accounts for 55% of Annex 1 emissions in 1990. With Russia's ratification of the Protocol in November 2004, the

A. Iacomelli (ed.), Renewable Energies for Central Asia Countries: Economic, Environmental and Social Impacts, 27–42.

number of Parties rose to 127, accounting for 61.6% of the Annex 1 1990 emissions. The Protocol entered into force in February 16, 2005.

Greenhouse gases covered under the Kyoto protocol are carbon dioxide, methane, nitrous oxide, hydrofluorocarbons, perfluorocarbons and sulfur hexafluoride. The Protocol requires Annex 1 countries to engage in domestic activities that reduce emissions or absorb emissions such as afforestation and reforestation. To help these countries reduce the costs of meeting their obligations, the Protocol establishes three flexible mechanisms, and these are: i) Emissions Trading (ET), ii) Joint Implementation (JI), and iii) Clean Development Mechanism (CDM). In Emissions Trading, Annex 1 countries are allowed to exchange among themselves parts of their assigned amount units (AAUs); in Joint Implementation, Annex 1 countries are allowed to claim emission reductions units (ERUs) arising from their investments in another Annex 1 country; and in Clean Development Mechanism, Annex 1 countries are allowed to claim certified emissions reductions (CERs) from their sustainable development investments in developing countries.

1 THE CLEAN DEVELOPMENT MECHANISM

The Clean Development Mechanism is a project-based mechanism where Annex 1 countries can purchase or claim CERs from projects implemented in developing countries (non Annex 1 countries) to be used for meeting their emissions reduction targets. Projects that qualify for CDM include the following: end-use energy efficiency, supply-side energy efficiency, renewable energy, fuel switching, agriculture, industrial processes, solvent and other product use, waste management, and sinks (afforestation and reforestation).

These projects must also satisfy two main conditions set by the protocol: additionality and contributions to sustainable development. The additionality condition states that projects must result in reductions in emissions that are additional to any that would occur in the absence of the project activity, and that the projects must lead to real, measurable and long term benefits. The sustainability condition states that projects must assist developing countries in achieving their sustainable development goals. There is however no guideline provided by the Protocol except that each country must develop its own criteria and assessment procedures.

The CDM process is governed by the CDM Executive Board. The Board accredits independent operational entities to validate proposed CDM projects, verify and certify emissions reductions. It also maintains a CDM registry, which issues CERs, manage the CER levy, and maintain CER account for each developing country hosting a CDM project. For developing countries to participate in CDM, the Protocol requires that each country must establish a national authority responsible for CDM, ratify the Kyoto Protocol and participate voluntarily.

Investors may prefer large-scale projects in developing countries since these projects generate large quantity of CERs at lower transaction cost per unit of emissions reduction. In order to remove the bias towards large-scale projects, the Executive Board has developed simplified modalities and procedures for small-scale projects to reduce transaction costs. Small-scale CDM projects are defined as the following: i) renewable energy project activities with a maximum output capacity equivalent of up to 15 MW; ii) energy efficiency improvement project activities which reduce energy consumption on the supply and/or demand side, by up to the equivalent of 15 GWh per year; or iii) other project activities that both reduce anthropogenic emissions by sources and directly emit less than 15 thousand tonnes of carbon dioxide equivalent annually.

Bundling of smaller projects to reduce transaction costs are also allowed by the Board as long as the total size of the bundled project satisfies the small-scale project criteria. On the other hand, de-bundling of a large-scale project into smaller scale projects is prohibited by the Board. A de-bundling test was developed by the Board as the following: small-scale projects are deemed de-bundled components of a large-scale project when the application to register another small-scale project shows that i) participants are the same entities to that of the previous project, ii) the project category and technology/measures remain the same, iii) the previous project was registered in the previous 2 years, and iv) the project boundary of the previous project is within 1 km of the proposed new project.

2 CARBON FINANCE AND PROJECT VIABILITY

CDM projects produce both conventional project output and carbon benefits (CERs). The value of carbon benefits and its impact on project viability are influenced by several factors such as the amount of CERs generated by the project, the price of CER and the transaction costs involved in securing CERs.

2.1 Quantity of CERs

The amount of CERs generated by the project depends on the greenhouse gas displaced by the project and the crediting period selected.

Renewable energy and energy efficiency projects displace carbon intensive electricity and/or heat generation. Grid-based or off-grid projects that displace more carbon intensive coal and diesel fuels generate more CERs than those that displace natural gas. Projects that capture methane and other greenhouse gases produce more CERs since the global warming potential (GWP) of methane and other gases are several times higher than that of carbon dioxide. Methane's GWP is 21 times, nitrous oxide is 310 times, hydrofluorocarbons (HFCs) range from 140-11,700 times, perfluorocarbons (PFCs) is on average 6,770 times and sulfur hexaflouride is 23,900 times higher than carbon dioxide.

The total CER generated is determined by the selected crediting period. The Marrakech accords specify two options for project developers: 7 years with twice the option of renewal (totalling 21 years) or, 10 years without renewal.

2.2 Price of CERs

The price of CERs is determined in the carbon market. The CER market is one of the fragmented carbon markets. The global carbon market consists of diverse greenhouse gas reduction transactions and can be broadly classified as follows: i) project-based or baseline and credit system. Emission reductions are created and traded through a given project or activity. CDM and JI are examples of the project-based system where CERs and ERUs are generated respectively; ii) allowance market or cap and trade system. Emission allowances are defined by regulations at the international, national, regional or firm level. Examples of allowance market include the Emissions Trading under the Kyoto Protocol (global), EU Emission Trading System or EU ETS (regional), the UK and the Danish trading systems (national), and BP and Shell internal trading (firm).

Most of market volume transactions are project-based, and the emissions reductions credits are intended either for Kyoto Protocol or non-Kyoto compliance. Buyers have various motives in engaging transactions in the global carbon market. Risk minimization objectives could be classified as follows: i) immediate compliance in the national markets where buyers seek to comply with existing legislative obligations and constraints; ii) Kyoto pre-compliance where buyers expect the project to be registered under either JI or CDM; iii) voluntary compliance where buyers aim to use the emission reductions to meet part of their voluntary targets; and iv) retail schemes where buyers wish to be climate-neutral in order to demonstrate their social responsibility or promote particular brand. In addition to risk minimization, other objectives include the following: i) learning by doing, ii) experimenting with diverse contract structures, iii) influencing policy, iv) broadening the envelope of flexibility, v) public relations, and vi) goodwill (PCF, 2003).

The fragmented nature of the global carbon market generates differentiated prices for emissions reductions as shown in Table 1. Allowance markets generate high emission reduction prices since the delivery risks are believed to be minimal. Though JI and CDM are both project-based, PCF pays higher prices for ERUs since JI are supported by Host Country Agreements and Assigned Amount Units, which reduces PCF's exposure to risks. ERUPT however in its January 2003 tender for JI projects have specified a price range similar to C-ERUPT tender for CDM projects.

Table 1. Carbon Emission Reduction Prices (per TCO2e)

Project-Based		**Allowance Markets**
Clean Development Mechanism	**Joint Implementation**	
PCF[1] • US$3.0-3.5 • premium of US$0.5 per ton of CO2e for projects with developmental components (Colombia Wind Farm) **C-ERUPT**[2] (maximum prices) • renewable energy – €5.5 • biomass energy - €4.4 • energy efficiency - €4.4 • fuel switch and methane - €3.3 • average price - €4.7[3] **Finish Government**[4] • small-scale - €2.47-3.2	**PCF**[5] • US$ 3.5-4.0 **ERUPT**[6] • First tender average price - €8.46 (closed in April 2001) • Second tender average price - €4.78 (closed in March 2002) • Third tender - expected price range - €3.0-5.0[7] (closed in January 2003) **Denmark-Romania JI**[8] • estimated price range €5.40-8.10	**Regional** • EU-ETS[8] €5.0-7.0 (indicative price); € 13.05[9] (forward price in Jan 2004); €7.17[10] (forward price in Apr 2004) **National** • UK-ETS[11] – Bid price £1.75, offer price £2.25 **Firm** • BP Emissions Trading Scheme[12] (Scheme discontinued in 2001) average in 2000 – US$7.6 average in 2001 – US$39.63

[1]PCF Annual Report 2002; [2]C-ERUPT Tender Document 2002; [3]Carbon Market Europe (March 21 2003); [4]http://global.finland.fi; [5]PCF Annual Report 2002; [6]Environmental Finance (February 2003); [7]GHG Market Trends 2/2003; Carbon Market Europe (March 7, 2003); [8]Carbon Market Europe (May 2 2003); [9]Evolution Markets LLC (Jan 2004); [10]Carbon Market Europe (April 15 2004); [11]Carbon Market Europe (August 15 2003); [12]ww.bp.com/files/15/Climate_Change_2001_performance_1541.pdf

The pricing of CER is highly speculative. The PCF considers several parameters in determining its price in the PCF's carbon purchase agreement. Moreover, certain project parameters command price premiums under the PCF program. These include: i) the existence of government guarantees, ii) project generation of social benefits, and iii) the exclusion of preparation costs in the total project cost. Among the CDM projects being contracted by PCF, a price premium of US$ 0.5 per TCO2e has been offered to the Colombia Jepirachi Wind Farm sponsors for the delivery of activities that improve the social conditions of the local indigenous population that hosts the project.

In C-ERUPT program, prices are also differentiated according to technology type. CER from renewable energy project forms the reference price (maximum price of EUR 5.5 per CER). CERs from sustainable grown biomass projects as well as from energy efficiency projects are priced 20% lower (maximum price of EUR 4.5) while those from fuel switching and methane recovery projects are 40% cheaper (maximum price of EUR 3.3).

At present there is no single CER price. It is differentiated according to risks, technology type and social development components. The current PCF CER rate ranges from US$3 to 4 per ton of CO2; under the C-ERUPT program, it revolves around US$ 4 to 4.5 per ton of CO2. The CER price differentiation could evolve into the following categories: i) CERs from projects that fulfil the WWF Gold Standard, ii) CERs from projects with community development features, iii) CERs from standard projects, and iv) long-term and temporary CERs from forestry projects (Michaelowa, A., CDM Monitor, March 11, 2004).

Several economic models forecast a single carbon price since these models assume a competitive and unfettered markets. With the US presence in the GHG market, these models projected a very high carbon prices. After the Bonn Agreement and Marakkech Accords, and with the absence of the US in the market, these models projected low carbon prices. In reality, the carbon markets are fragmented and prices generated by these markets are differentiated. In a recent GHG market analysis, Natsource (2002) forecasts prices for project-based carbon emission reductions (both JI and CDM markets) to vary from US$3 to 5 for the period 2002-2005, US$2.5 to 9.0 during 2005-2007, and US$5 to11 from 2008-2012.

2.3 Transaction costs

Transaction costs are those that arise from initiating and completing transactions to secure CERs. These consist of pre-operational costs (or upfront costs), implementation costs (i.e. costs spread out over the entire crediting period), and trading costs (Table 2). Pre-operational costs include direct expenses for search, negotiation, validation, and approval. Implementation costs are those incurred for monitoring, certification, and enforcement while trading costs are those incurred in trading CERs such as brokerage costs and costs to hold an account in national registry.

PCF's pre-operational transaction costs amounts 229 thousand Euros (265 thousand dollars) while Ecosecurities (2002) estimates the minimum up-front transaction cost at around 70 thousand Euros (£42,000) (Table 3).

The CDM Executive Board has recently determined the registration fee for CDM projects. Fees for small-scale projects are fixed at US$5,000 while those for large-scale projects are between US$10,000 – 30,000. The registration fees are the following: i) ≤ 5,000 annual CO2 equivalent reduction - US$ 5,000; ii) >15,000 and ≤ 50,000 annual CO2 equivalent reduction - US$ 10,000; iii) >50,000 and ≤ 100,000 annual CO2 equivalent reduction - US$ 15,000; iv) >100,000 and ≤ 200,000 annual CO2 equivalent reduction - US$ 20,000; and v) >100,000 annual CO2 equivalent reduction - US$ 30,000. This rate also applies for bundled projects.

Several studies show that the transaction cost per ton of CO2 for large projects is very small or even negligible while that for small-scale projects is quite significant. Given this, it is obvious that investors would prefer large-scale projects. Fast-tracking small-scale projects (simplifying the procedures and standardizing the information and reporting requirements) not only reduces transaction costs but also improves project financial viability. According to Ecosecurities (2002), fast-tracked procedures lead up to around 67% reduction in transaction costs.

Table 2. CDM Transaction Costs

	Transaction Cost Component	Description
Pre-implementation phase	Search Costs	Costs incurred by investors and hosts as they seek out partners for mutually advantageous projects
	Negotiation Costs	Includes those costs incurred in the preparation of the Project Design Document that also documents assignment and scheduling of benefits over the project time period. It also includes expenses in organizing public consultation with key stakeholders.
	Baseline determination	Development of a baseline
	Approval costs	Costs of authorization from host country
	Validation Costs	Costs incurred in reviewing and revising the Project Design Document by operational entity
	Review Costs	Costs of reviewing a validation document
	Registration Costs	Registration by UNFCCC Executive Board/JI Supervisory Committee
Implementation Phase	Monitoring Costs	Costs to collect data
	Verification Costs	Costs to hire an operational entity and to report to the UNFCCC Executive Board/Supervisory Committee
	Review Costs	Costs of reviewing a verification
	Certification Costs	Includes costs in the issuance of Certified Emission Reductions (CERs for CDM) and Emission Reduction Units (ERUs for JI) by UNFCCC Executive Board
	Enforcement costs	Includes administrative and legal costs incurred in enforcing transaction agreements
Trading	Transfer Costs	Brokerage costs
	Registration Costs	Costs to hold an account in national registry

Source: Michaelowa, A., Stronzik, M., Eckerman, F., and Hunt, Alistair, 2003.

Table 3. CDM Transaction Cost Estimates

	Project Cycle	EcoSecurities, 2002 (£)	PCF (US $)
Pre-operational Phase Design	Preparation and review		40,000
	Baseline Study	12,000 – 15,000	20,000
	Monitoring Plan	5,000 – 10,000	20,000
	Environmental Assessment	-	
	Stakeholder Consultation	-	
	Approval	-	
	Validation	10,000 – 20,000	30,000
	Consultation and project appraisal		105,000
	Legal and Contractual Arrangements	15,000 – 25,000	50,000
Operational Phase	Sales of CERs	5% - 15% of CER Value	
	Adaptation Levy[1]	2% of the CER value annually	
	Risk Mitigation	1%-3% of CER value annually	
	Verification	5,000 per audit	25,000 (initial) 10,000-25,000 (periodic) 10,000-20,000 (periodic supervision)
	Executive Board Administration	To be determined (X% of CER value)	

[1] Projects in least developed countries are exempted from the 2% adaptation levy.
Sources: Ecosecurities, 2002; PCF Annual Report 2002.

2.4 Impact of CERs on project feasibility

The net financial gain derived from the sale of CERs is the difference between the project CER value and the transaction costs. Three elements influence the net impact of CERs on project profitability: value of CERs (low CER value implies low net benefits), overall transaction costs (high transaction costs yield low net benefits), and up-front transaction costs (high upfront payments could also result in low benefits). Project developers generally expect *up-front* transaction costs within the range of 5 to 7% of the net present value of the revenue or *total* transaction costs around 10 to 12% of the net present value of revenue (Ecosecurities, 2002).

A positive net financial gain means that CER revenues improve the financial viability of the project. Table 4 presents the impacts of carbon financing to the proposed 60 MW Wind Farm project in Zafarana, Egypt. For the CER price scenarios of US$3 and 10 per ton of CO2 equivalent, the project's net present value increases by 173% and 588% respectively. The project's internal rate of return increases by 1.04 and 3.38 percentage points while the return of equity rises by 2.73 and 8.24 percentage points for the respective CER price scenarios.

Table 5 shows the impact of CERs on IRRs in selected projects. The effect of CER cash flow on project IRRs vary by project type. The impact of CERs on wind power project IRR is relatively small (few percentage points increase) while it is substantially important for fugitive methane capture projects. More CERs are generated by methane capture projects since the global warming potential of methane is 21 times higher than carbon dioxide. This makes methane capture projects relatively attractive to CDM project developers. In fact, for the first 45 projects submitted to the CDM Executive Board for methodology review, 27% (12 projects) are methane gas capture projects.

Table 4. Impact of carbon financing on the proposed 60-MW Zafarana Wind Farm Project in Egypt

Economic Indicators	Without carbon finance	With carbon finance	
		US$3 per ton CO2eq	US$10 per ton CO2eq
Internal Rate of Return	5.63%	6.67%	9.01%
Net Present Value	US$2,954,117	US$8,065,191	US$20,320,777
Return on Equity after taxes	19.10%	21.83%	27.34%

Note: Financial and economic data are given in Appendix 3.1
Source: Ringius, L., Grohnheit, P.E., Nielsen, L.H., Olivier, A., Painuly, J., and Villavicencio, A. 2002. Wind Power Projects in the CDM: Methodologies and Tools for Baselines, Carbon Financing and Sustainability Analysis. Risoe National Laboratory.

Table 5. Impact of CERs on project IRR

Country	Project	IRR without carbon finance (%)	IRR without carbon finance (%)	Change in IRR (%)
Costa Rica	wind power	9.7	10.6	0.9
Jamaica	wind power	17.0	18.0	1.0
Morocco	wind power	12.7	14.0	1.3
Chile	Hydro	9.2	10.4	1.2
Costa Rica	Hydro	7.1	9.7	2.6
Guyana	Bagasse	7.2	7.7	0.5
Brazil	Biomass	8.3	13.5	5.2
India	solid waste	13.8	18.7	5.0

Source: PCF Annual Report 2001

3 CER TRANSACTIONS

Project-based transactions dominate the global trade of greenhouse gas emission reductions. It represented 85% of the total transaction volume in 2002, and 97% between 1996 and 2002 (PCF Plus, 2002). The total carbon market volume traded in 2001 was about 13 MMTCO2e, increasing to 29 MMTCO2e in 2002, and reaching to more than 70 MMTCO2e in the first 10 months of 2003 (PCF, 2003). Point Carbon (2004) projected that the total volume will reach 100 MMTCO2e in 2004.

The World Bank's PCF and the Dutch Government's C-ERUPT tender are the current main buyers of CERs through direct purchase transactions. As of 2003, the PCF has signed 7 emission reductions purchase agreements (ERPAs) with total emissions reductions of 12.19 MMTCO2e. Also, PCF has 144 projects under preparation and received 420 project idea notes. The C-ERUPT tender, on the other hand, approved 18 projects in 2003 aimed to generate emissions reductions of 16.7 MMTCO2e. A number of PCF projects have been operational since 2002. Most of the PCF and C-ERUPT's projects would be commissioned between 2003-2007. CDM portfolios were also launched by Austria, Denmark, Finland and Sweden. The Austrian government opened a tender for CDM projects in December 2003. Denmark is cooperating with Thai industries and will select 5 projects for actual CDM implementation. The Finnish Government launched a tender for small-scale projects and is currently engaged with 7 CDM projects. Sweden launched a tender in 2002 and selected 5 projects in India, Brazil and South Africa. Most recently, Belgium announced its plan to purchase emissions reductions of around 2.46 MMTCO2e annually in the period 2008-2012.

CER procurement funds are growing and expanding. As shown in Table 6, three new public-private partnership funds have been recently launched by the World Bank: the Community Development Carbon Fund, the Bio-Carbon Fund and the Italian Carbon Fund. Public-private partnership funds to purchase CERs were also established by the European Investment Bank, Japanese Banks, Germany's KfW and Ecosecurities-Standard Bank of London (Danish CDM Facility). In addition to the Government of the Netherlands, several European governments have launched CDM funds. These governments have used several vehicles in CER procurement such as government-own tenders through banks and multilateral institutions. Bilateral transactions are also emerging. Several European governments and the government of Canada have signed MOUs with several Latin American and Asian countries for the development of projects and supply of CERs.

CERs purchased through public-private partnership and government funds are mainly used for Kyoto compliance. Private funds are also being established to secure CERs for purposes other than compliance. Mitsubishi Corporation of Japan recently purchased emission rights from a Chilean Hydropower project for trading purposes. More recently, Cumbria Energy, Investic Bank and Less carbon launched ICECAP, a vehicle to purchase CERs for large industrial emitters and Annex 1 governments. Mitsubishi Securities Company and Mizuho Securities Company are also planning to be involved in the purchase of carbon emissions certificates to cater to the needs of their business clients. Aside from trading, institutions interested in becoming carbon neutral with their activities could be another buyer of CERs. The Dutch Development Finance Company, for example, have announced their intention to compensate carbon dioxide emissions from their activities in developing countries and that it plans to purchase CERs from projects the company finances.

Table 6. CER Procurement Funds

Public-Private Partnerships	***Government Funds***	***Private Funds***
Multilateral Institutions *The World Bank* □ Prototype Carbon Fund (US$ 180 million) □ Community Development Carbon Fund (US$ 100 million) □ World Bank Bio-Carbon Fund (US$ 100 million) □ Italian Carbon Fund (US$15 million) □ Spanish Carbon Fund (under discussion) *European Investment Bank* □ Proposed Carbon Investment Trust **Other Financial Institutions** **Japan** *Japan Bank for International Cooperation (JBIC) and Development Bank of Japan* □ Joint Carbon Fund (10 billion yen) **Germany** *KfW* □ German Carbon Fund (€ 50 million) **Denmark** *Ecosecurities and Standard Bank of London* □ Denmark Carbon Facility (DKK 59 million)	**Own Tender** □ Denmark CDM Program □ Dutch Government C-ERUPT Program □ Finnish CDM/JI Pilot Program (€ 20 million) □ Sweden International Climate Investment Program - CDM □ Austria JI/CDM Procurement Program □ Belgium CDM/JI Program **Through Commercial/Development Banks** □ Rabo Bank (Dutch Government) Through Multilateral Institutions □ World Bank (The Netherlands Clean Development Facility - € 70 million) □ IFC (IFC-Netherlands Carbon Facility - € 44 million) **Through Bilateral Transactions (signed MOUs)** □ Austria: discussions with China □ Canada: Costa Rica, Colombia, Chile, Nicaragua, Tunisia, South Korea □ Denmark: Malaysia, Chile; discussions with China, South Africa □ Finland: China, Costa Rica, El Salvador, Nicaragua, India □ France: Colombia and Morocco □ Italy: Algeria, China, Cuba, Cyprus, Egypt, El Salvador, Israel, Morocco □ Netherlands: Colombia, Costa Rica, El Salvador, Panama, Uruguay, Bolivia, Nicaragua, Guatemala, Honduras. Under negotiation: Indonesia, Philippines).	**For trading** □ ICECAP (Cumbria Energy, Investec Bank and Less Carbon) □ Mitsubishi Corporation (purchased emission rights from Hidroelectrica Guardia Vieja, SA) □ Mitsubishi Securities Co. □ Mizuho Securities Co. **Voluntary use (carbon dioxide neutral)** □ Dutch Development Finance Company

4 SECURING PROJECT FUNDS

CDM projects require upfront investments that are normally obtained from different sources such as loans, equity, grants, and upfront payments for emission reductions. Loans or debts refer to funds lent to CDM project owners by financiers. Debt can be obtained through public markets (bonds) or private placements (bank loans and institutional debt). Equity[i] refers to funds funneled to the CDM project by company shareholders. Equity may be sourced from internal sources (sponsors) or external investors (public or private markets). The return on equity is obtained either from dividends or from sale of shares. Grants[ii] are funds provided by institutions and governments to CDM project owners and developers who contribute to donors' objectives. Grants need not be repaid and oftentimes, cover only a percentage of project costs. The carbon purchase agreement often stipulates payment on agreed price upon delivery of CERs but CER buyers sometimes provide upfront payment upon purchase. For example, the PCF provides upfront payment up to 25% of the total CER value. However, to compensate for increased risk, upfront payments are discounted.

Like conventional projects, financing CDM projects can be arranged either through corporate or project financing. These are described as follows: i) in project financing, a project company is formed and investments are viewed as assets of the company. Investment funds are sourced either from equity or debt. Assets and cash flow secure debts. Creditors do not have recourse to the other resources of sponsors; ii) under corporate financing, new projects are undertaken as extension of assets of the existing company. Capital investments and borrowing are not placed under the project account. Loans are considered as company debts and lenders have full recourse to all the assets and revenues of the company over and above those generated in the new project.

Additional project revenues such as CERs could leverage debt financing in project financing arrangement and could be used to service debts. Guest et al (2003) presents that the carbon cash flow can help increase debt carrying capacity. The carbon revenues could help increase debt leverage of project by increasing the debt service coverage ratio (DSCR) levels of the project. In addition to improving debt capacity, there are other options to debt service through the carbon cash flow. These include: pre-paying debt based on Forward Emission Reduction Purchase Agreements (ERPAs); depositing carbon cash flow directly with banks for credit against debt service thereby lowering liability on electricity cash flow; and using ERPAs and/or forward carbon sales as collateral for loans (this is the case for Plantar project in Brazil where the CER purchase agreement with the PCF was used as collateral for commercial bank financing).

5. CONCLUSION

The Clean Development Mechanism (CDM) stimulates investments on renewable energy projects in developing countries. The carbon asset in the form of certified emission reductions (CERs) generated by CDM projects improves project viability and attracts capital to finance the development of these projects. The entry into force of the Kyoto Protocol creates real demand for CERs from Annex 1 countries and the current policies of many Annex 1 countries, particularly European countries, to supplement emissions reductions from domestic actions with CERs from projects in developing countries result in the creation of carbon funds dedicated to CER procurement. These funds, in turn, leverage equity and debt financing that are necessary to develop the project.

REFERENCES

BP Emissions Trading Scheme. http://www.bp.com
Carbon Market Europe. http://www.pointcarbon.com
CERUPT Tender Document, 2002. http://www.senter.nl
CDM Monitor. http://www.pointcarbon.com
Environmental Finance. vol 4, no. 4, February 2003.
Evolution Markets LLC. http://evomarkets.com
EcoSecurities, Clean Development Mechanism (CDM): Simplified Modalities and Procedures for Small-scale Projects, A DFID Report, May 2002. http://www.ecosecurities.com
Guest, Justin, Stuart, Marc and Wellington, Fred. The Role of Emissions trading in Asian clean energy finance, JASSA 2003; issue 4.
GHG Market Trends. http://www.ghgmarket.info
Lee, M. K., Fenhann, J., Haelsnæs, K., Pacudan, R., Olhoff, A., *CDM Information and Guidebook*, UNEP Risoe Center, Denmark, June 2004.
Lee, M.K. and Pacudan, R. Capacity Building for CDM, Asia Pacific Technology Monitor, March-Apr 2003, pp. 29-32.
Michaelowa, Axel, Stronzik Marcus, Eckerman, Fraucke and Hunt Alistair, Transaction Cost of the Kyoto Mechanisms, *Climate Policy* 2003; 3:261-278.
Pacudan, R. and Lee, M.K. Overview of the CDM market and CER prices, Responding to Climate Change 2003, pp. 72-75.
Prototype Carbon Fund. Annual Reports 2001, 2002, 2003.
http://www.prototypecarbonfund.org
Prototype Carbon Fund. State and Trends in the Market 2004.
http://www.prototypecarbonfund.org
Ringius, L., Grohnheit, P.E., Nielsen, L.H., Olivier, A., Painuly, J., and Villavicencio, A. Wind Power Projects in the CDM: Methodologies and Tools for Baselines, Carbon Financing and Sustainability Analysis. Risoe National Laboratory, 2002.

United Nations Development Programme. The Clean Development Mechanism: A User's Guide. 2003.
United Nations Convention Framework Convention on Climate Change. http://unfccc.int

---

[i] Equity fund providers that target carbon credits include: Dexia-FondElec Energy Efficiency and Emission Reduction Fund (71 million Euros, since 2000). FondElec Latin American Clean Energy Services Fund (US$ 31 million, since 2001). Global-Asia Clean Energy Services Fund, FE Clean Energy Group (US$100-150 million) (seeking for 20-25 % returns). Private-public partnerships that provide upfront financing to CDM projects include Climate Investment Partnership.

[ii] The Danish Government offers grants to firms in Thailand to kick start CDM projects. In addition, The European Investment Bank intends to launch a Transaction Assistance Facility which will help in project identification and preparation and carbon credit marketing. The facility will provide a grant, which is repayable from the revenue generated by the sale of carbon.

Chapter 4

INTERNATIONAL CO-OPERATION ON ENERGY TECHNOLOGIES RESEARCH AND DEVELOPMENT

The International Energy Agency Framework

Alicia Mignone*

Energy and Science Advisor,
Permanent Delegation of Italy to OECD

1. INTRODUCTION

1.1 The IEA

The International Energy Agency (IEA) is an autonomous body which was established in November 1974 within the Framework of the Organization for Economic Co-operation and Development (OECD) to implement an International Energy Program in the wake of the first oil shock. It carries out a comprehensive program of long term co-operation on energy among twenty-six[1] of the OECD thirty member countries. The initial objectives were to represent the major consuming nations and to work for stability in the world energy markets.

The basic aims of the IEA are:

- To maintain and improve systems for coping with oil disruptions;

[1] IEA Member countries: Australia, Austria, Belgium, Canada, the Czech republic, Denmark, Finland, France, Germany, Greece, Hungary, Ireland, Italy, Japan, the Republic of Korea, Luxembourg, the Netherlands, New Zealand, Norway, Portugal, Spain, Sweden, Switzerland, Turkey, the United Kingdom, the United States. The European Commisions also takes part in the work of the IEA.

A. Iacomelli (ed.), Renewable Energies for Central Asia Countries: Economic, Environmental and Social Impacts, 43–58.

- To promote rational energy policies in a global context;
- To operate a permanent information system on the international oil market;
- To improve the world's energy supply and demand structure;
- To assist in the integration of environmental and energy policies.

The shared goals adopted by the IEA Ministers at their June 3rd meting in Paris 1993 can be summarized by the so-called IEA "3Es": Energy Security, Environmental Protection and Economic Growth.

The development and deployment of cleaner, more efficient energy technologies are fundamental requirements within any strategy to fulfill the 3Es. The text approved by the Ministers more than ten years ago, still valid and relevant, contains a clear message on the need of clean energy technologies: [2]

- "More environmentally acceptable energy sources need to be encouraged and developed. Clean and efficient use of fossil fuels is essential. The development of economic non-fossil sources is also a priority. A number of IEA members wish to retain and improve the nuclear option for the future, at the highest available safety standards, because nuclear energy does not emit carbon dioxide. Renewable sources will also have an increasingly important contribution to make."
- "Continued research, development and market deployment of new and improved energy technologies make a critical contribution to achieving the objectives outlined above. Energy technology policies should complement broader energy policies. International co-operation in the development and dissemination of energy technologies, including industry participation and co-operation with non-Member countries, should be encouraged."

1.2 The organization of the IEA

The fig. 1 shows the organization of the IEA: the blocks represent the Governing Board, the maximum decision making body; the Secretariat, comprising experts from Member countries and the three Standing Groups and two Committees, that benefit from the participation of delegates and experts from Member Countries.

[2] For the complete version of the shared goals see: http://www.iea.org/dbtw-wpd/Textbase/about/sharedgoals.htm

Figure 1. The organization of the IEA

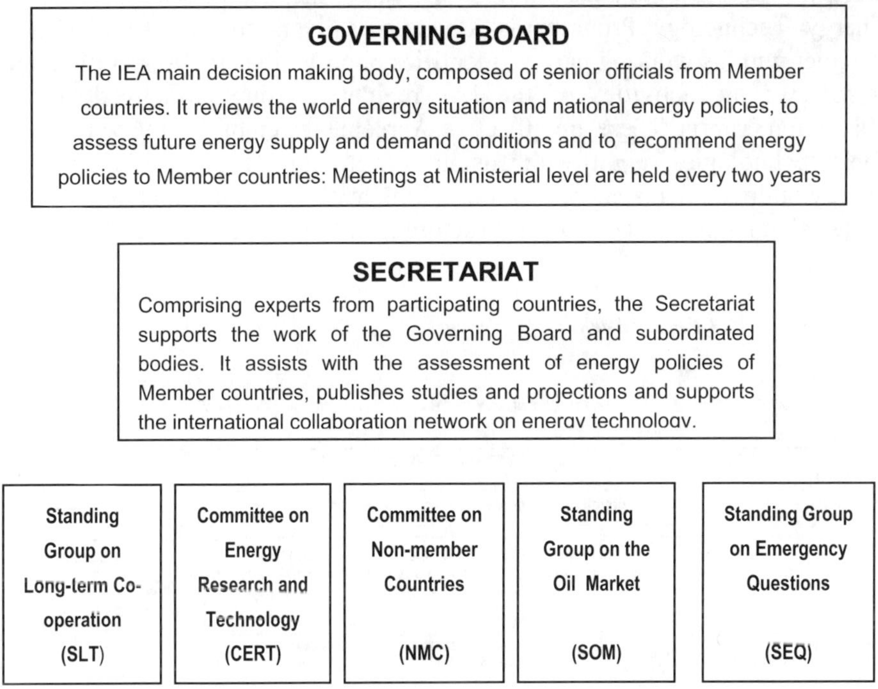

2. THE COMMITTEE ON ENERGY RESEARCH AND TECHNOLOGY (CERT)

The Committee on Energy Research and Technology established in the 1975 has the mandate to promote the research, development and deployment (RD&D) of clean energy technologies through international networking, co-operation, collaboration, analysis and policy advice. The CERT objectives as established in the "Strategic Plan" [3] are to encourage cost-effective energy technology collaboration; to produce high-quality energy technology policy analysis; to cooperate with national governments, the OECD and other international organizations and to keep the Governing Board informed on its activities and progress.

A major component of the IEA's International Energy Technology Co-operation Program are the Implementing Agreements (IAs)

[3] IEA, The First 30 years 1974-2004, Vol. 4, OECD/IEA 2004, Paris

that focus on fossil fuels, renewable energy, efficient energy end-use and fusion power. Effective dissemination of results and findings is a crucial part of each program's mandate and in some cases, the sole activity. The IEA's Energy Technology Program as a whole is overseen by CERT. CERT together with its working parties, identifies opportunities for cooperation and it reviews the activities of the IAs regularly, using an agreed set of evaluation criteria. There are 40 active Agreements, eight dealing with fossil fuels technologies, eight with fusion power, eight related to renewable energy, thirteen with energy end-use, one with hydrogen and four inter-sectorial agreements. The CERT structure is presented in figure 2.

Figure 2. The IEA Energy Technology Network

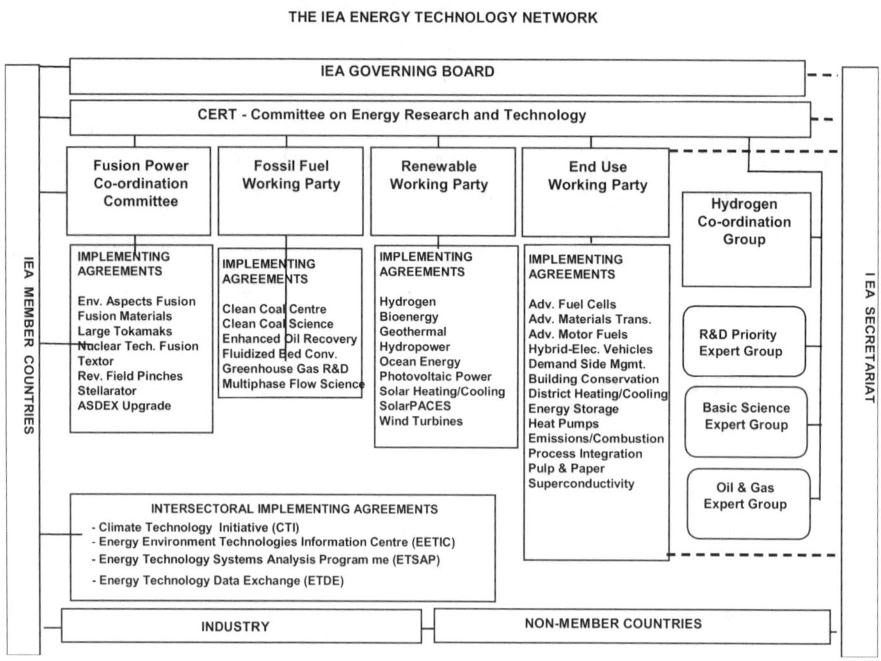

3. THE IMPLEMENTING AGREEMENTS

The legal framework for the RD&D international collaboration is constituted by the Implementing Agreements that offer a flexible mechanism which accommodates various forms of energy technology co-operation. The Implementing Agreements are a contractual relationship establishing the rights and the obligations of the participants and the management structure to guide the work under each Implementing Agreement.

3.1 Benefits

The benefits of the participation to the Implementing Agreements are analogous to those offered by the participation to international projects and partnerships. They can be summarized as follows: to share costs and to pool technical resources; to have access to a network of researchers and to avoid duplication of efforts; to reinforce national R&D capabilities; to accelerate technology development and deployment; to better disseminate information on improved energy technologies and to boost trade and exports. In addition to the normal benefits of collaboration, it is worth noting that the IEA links Europe, North America, Japan and Australia and increasingly other non Member countries are participating to the R&D activities, e.g. Russia, China, Brazil and South Africa. Moreover, the Agreements involve governments and industrial partners, offering intellectual property right protection, equal voluntary partnership and flexibility.

3.2 Rules that apply

The rules that apply are the IEA Framework for International Energy Technology Collaboration [4]; the decisions of the IEA Governing Board and of the Committee on Energy Research and Technology and the decisions of the Executive Committee of the Implementing Agreement.

The framework has been approved by the IEA Governing Board on April 3rd 2003 in replacement of the Guiding Principles for Co-operation in the field of Energy Research and Development, aiming at streamlining the structure of the IAs and at better responding to the realities and the needs of today energy sector.

[4] see www.iea.org/textbase/techno/index.asp

3.3 How activities are funded

The activities can be funded following different modalities[5], the more important being cost or task sharing, respectively. In the cost sharing option, the participants contribute to a common fund; the project is contracted out to a specific entity and the results are provided to all. In the task sharing modality, the participants devote specified resources and personnel. The project may be a common work program or just information sharing.

3.4 Implementing Agreement - Players

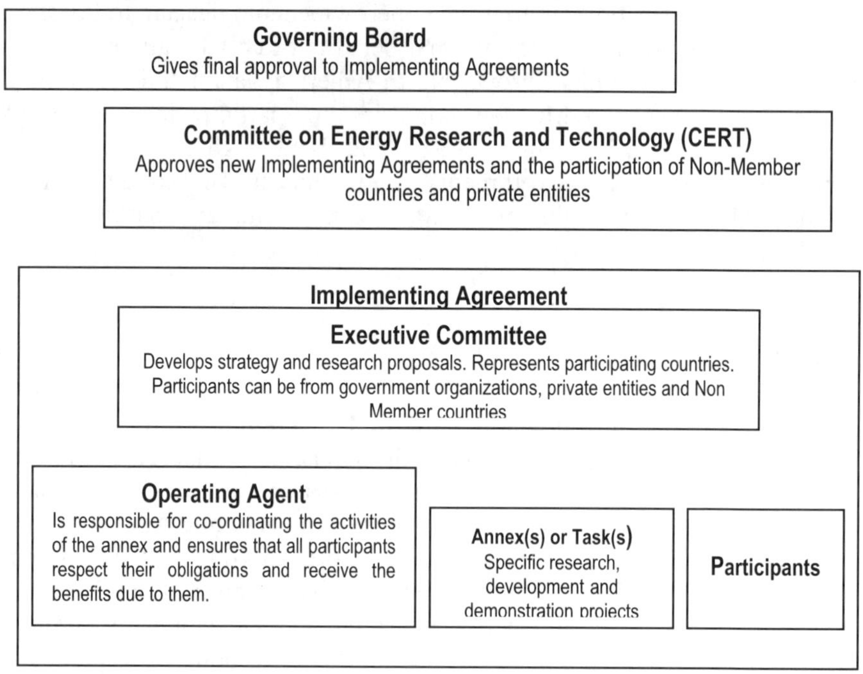

The Executive Committee (Ex-Co) defines the terms and conditions of Contracting Parties and Sponsors, namely with regard to their respective rights and obligations, subject to certain limitations set out in the Framework; it defines the program of work; approves the budget and the

[5] For more details. see "Overview of Financing modalities for selected Implementing Agreements" by Peter Tulej (peter.tulej@iea.org)

management structure of the Implementing Agreement. The Ex-Co verifies that the financial and/or in-kind contribution of Participants respects the principle of equitable sharing and that an adequate protection of the intellectual property rights is adopted. The Sponsors are from several OECD member countries and non Member countries.

3.5 Admission to an Implementing Agreement

The minimum requirements set out by the Framework are the following ones:

1. Unanimous decision by the Ex-Co to invite a government (Contracting Party: CP) or a different entity (Sponsor) to join the Implementing Agreement;
2. Acceptance of the terms and conditions of participation by the CP or Sponsor (for the admission of Sponsors, the CERT approval is also needed) and
3. Signature of the actual CP or Sponsor.

4. CURRENT IMPLEMENTING AGREEMENTS

The forty active Implementing Agreements deal with renewable energy; energy technology modeling; efficient end-use technologies; fossil fuel technologies and nuclear fusion science and technology. There are approximately ninety tasks or annexes and ten countries per agreement. The Participants belong to all OECD member countries and from eleven Non-Member countries. The Non-Member countries are Algeria that participates to two Implementing Agreement; Brazil, to 3; China to 3; Croatia to 1 ; Egypt to 1; Israel to 4; Lithuania to 1; the Russian Federation to 7; South Africa to 2; Ukraine to 1 and Venezuela to 2.

5. IMPLEMENTING AGREEMENTS ON RENEWABLE ENERGY TECHNOLOGIES

There are eight Implementing Agreements that deal with renewable energy technologies. The subjects treated are: bio-energy; geothermal energy; hydropower; photovoltaic power systems; solar heating and cooling; solar power and chemical energy systems; wind energy and ocean energy systems. The main activities of the IAs as well as the participating countries are described below.

5.1 Bioenergy[6]

The bioenergy IA was set up in 1978, has twelve active tasks and counts with the participation of twenty countries, including two Non-member countries, plus the European Commission. The countries are the following ones: Australia, Austria, Belgium, Brazil, Canada, Croatia, Denmark, European Commission, Finland, France, Ireland, Italy, Japan, New Zealand, Norway, South Africa, Sweden, Switzerland, The Netherlands, United Kingdom.

IA on Bioenergy aims to accelerate the use of environmentally sound and cost-competitive bioenergy on sustainable basis, and thereby achieve a substantial contribution to future energy demands. The scope of the work is to integrate research themes across the value chain.

The tasks deal with the Socio-economic drivers in Implementing Bioenergy Projects; Short rotation crops for bioenergy; Biomass production for energy from sustainable forestry; Biomass combustion and co-firing; Thermal gasification of biomass; Pyrolysis of biomass; Techno-economic assessments for Bioenergy applications; Energy from integrated solid waste management systems; Energy from biogas and landfill gas; Greenhouse Gas balances of biomass and bioenergy systems; Liquid biofuels from biomass; Sustainable international bioenergy trade: Securing supply and demand; Bioenergy systems analysis.

5.2 Geothermal Energy[7]

The Implementing Agreement on Geothermal Energy commenced in March 1997. The objectives are the exchange of information, the common development of new technologies, and the dissemination of information on the environmental advantages of geothermal energy. Work underway includes identification of, and the development of means to avoid or minimize, adverse environmental impacts that can arise from the use of geothermal energy; the development of hot dry rocks and other technologies for commercial heat extraction; and the commercial development of deep geothermal resources.

[6] For more information, consult the website: www.ieabioenergy.com

[7] Website: www.iea-gia.org

The IA has five active tasks and ten participating countries plus the European Commission. The countries are Australia, Germany, Greece, Iceland, Italy, Japan, Mexico, New Zealand, Switzerland and the United States of America.

The annex on Environmental Impacts of Geothermal Energy Development is devoted to encourage sustainable development of geothermal energy resources, quantify adverse or beneficial impacts and identify means of avoiding, remedying or mitigating adverse effects.

The annex projects on Enhanced Geothermal Systems aim to address new and improved technologies to artificially simulate a geothermal resource to enable commercial heat extraction. The annex on Deep Geothermal Resources addresses the issues necessary for the commercial development of deep geothermal resources which prevail at depths of approximately 3,000 meters and deeper.

The task on Advanced Geothermal Drilling Techniques pursues advanced geothermal drilling research and investigate all aspects of well construction.

The objectives of the Direct Use Annex are: to define and characterize the direct use applications for geothermal energy, with emphasis on defining barriers to widespread application; to identify and promote opportunities for new and innovative applications; to define and initiate research to remove barriers, to enhance technology and economics, and to promote implementation tests, to standardize equipment and to develop engineering standards.

5.3 Hydropower[8]

The IEA Hydropower Agreement is a working group of governments and industry which intends to provide objective, balanced information about the advantages and disadvantages of hydropower. It has five active tasks and three under development with nine participating countries, including one Non Member country, China. The other eight are Canada, Finland, France, Italy, Japan, Norway, Sweden and the United Kingdom.

The annexes deal with the following themes:

- The Hydropower Risk Management is aimed at bringing together expertise from participating countries on the subject of upgrading existing installations with capacities of more than 10MW, i.e. intermediate to large size.

[8] Reference: www.ieahydro.org

- The task on Small-scale hydropower addresses technological, organizational and regulatory issues related to small hydro projects (less than 10MW and more than 50KW).
- The purpose of the annex on Public awareness is to increase global understanding of the current and future rolls and importance of hydropower in the global energy portfolio.
- The objective of the Hydropower competence network for educational training is to create a pilot version of an international network for training of personnel in the hydropower industry.
- Finally the task on Hydropower good practices aims to develop training materials in the areas of planning, operations and maintenance of hydropower installations, making use of latest information technologies to disseminate its results.

5.4 Ocean energy systems[9]

The Implementing Agreement on Ocean Energy Systems commenced in October 2001. There are six participating countries plus the European Commission: Canada, Denmark, Ireland, Japan, Portugal and the United Kingdom.

The Agreement's mission is to enhance international collaboration to make ocean energy technologies a significant energy option in the mid-term future. Through the promotion of research, development, demonstration and information exchange and dissemination, the Agreement's objective is to lead to the deployment and commercialization of Ocean Energy Technologies.

Current priorities are ocean waves and marine current systems developed in two tasks. One task is devoted to collate, review and facilitate the exchange and dissemination of information on the technical, economic, environmental and social aspects of ocean energy systems. The other task aims to the Development of recommended practices for testing and evaluating ocean energy systems and, in this way, improve the comparability of experimental results.

[9] for reference: www.iea-oceans.org

5.5 Photovoltaic Power Systems [10]

The Photovoltaic Power Systems (PVPS) Program is a collaborative R&D Agreement conducting projects on the application of solar photovoltaic electricity since 1992. IEA - PVPS operates worldwide via a network of national teams in member countries. There are twenty-one participating countries plus the European Commission: Australia, Austria, Canada, Denmark, Finland, France, Germany, Italy, Japan, Korea, Mexico, Netherlands, Norway, Portugal, Spain, Sweden, Switzerland, Turkey, the United Kingdom, the USA.

The mission of the PVPS program is "To enhance the international collaboration efforts through which photovoltaic solar energy becomes a significant renewable energy source in the near future."

The underlying assumption is that the market for PV systems will gradually expand from the present niche markets of remote applications and consumer products, to the utility market, through building-integrated and other diffused and centralized PV generation systems. This market expansion requires the availability of and access to reliable information on the performance of PV systems, design guidelines, planning methods, etc. to be shared with the various actors defined above.

The PVPS program aims to realize its mission by adopting the following objectives related to reliable PV power system applications for the target groups: utilities, energy service providers and other public and private users: to contribute to the cost reduction of their applications; to increase the awareness of their potential and value; to foster their market deployment by removing technical and non-technical barriers; to enhance technology co-operation with non-IEA countries.

The activities (eight tasks) focus on the following subjects: the exchange and dissemination of information about solar electricity; the operational performance, maintenance and sizing of photovoltaic power systems; the use of PVPS in stand-alone and island applications; the design and grid interconnection of building integrated and other dispersed PV systems; PVPS in the built environment; very large scale PV power generation systems in remote areas; cooperation with developing countries and upcoming urban scale grid-connected PV applications.

[10] Website: www.iea-pvps.org

5.6 Solar Heating and Cooling [11]

The Solar Heating and Cooling Implementing Agreement was one of the first collaborative R&D programmes to be established within the IEA, and, since 1977, its participants have been conducting a variety of joint projects in active solar, passive solar and photovoltaic technologies, primarily for building applications.

There are twenty participants and the European Commission: Australia, Austria, Belgium, Canada, Denmark, Finland, France, Germany, Italy, Japan, Mexico, Netherlands, New Zealand, Norway, Portugal, Spain, Sweden, Switzerland, the United Kingdom and the USA.

A total of thirty-one Tasks (projects) have been undertaken since the beginning of the Solar Heating and Cooling Programme.

In 2002, the Solar Heating and Cooling Programme celebrated 25 years of accelerating the solar market and contributing to the R&D of solar technologies. The activities undertaken by the Programme ranged from producing reports on the solar thermal collector market in IEA countries and on solar energy activities in IEA countries to starting new work in the areas of storage, industrial process heat, and building energy analysis tools.

The Tasks which were active in 2002 dealt with:

- Building energy analysis tool: aimed to creating standard methods of testing building energy analysis software, which supports implementation of national energy standards and use of renewable energy.
- Active solar procurement: to form buyer groups to purchase technologies, collect and analyse information on procurement, and to create tools to facilitate it.
- Solar assisted air conditioning buildings: to define performance criteria, develop methods for systems integration, develop technologies for solar assisted cooling systems.
- Solar combined systems: to collect international expert knowledge in the field of combined domestic hot water and space heating and develop and optimise the system technique.
- Performance of Solar Façade Components: to determine the performance of materials and components, such as advanced glazing; and to promote increased their use by developing and applying appropriate methods for assessment of durability, reliability and environmental impact.

[11] Solar Heating and Cooling Program website: www.iea-shc.org

- Sustainable Solar Housing : To address cost optimisation of the mix of concepts reducing energy losses, increasing useable solar gains and efficiently providing backup in order to achieve the same high performance.
- Solar Crop Drying: To address the barriers by providing technical and commercial information and experience gained from the design, construction and operation of full working demonstration systems for a variety of crops and a number of geographical regions.
- Daylighting in Buildings in the 21st Century: To make daylighting the typical and preferred design solution for lighting buildings in the 21st century by integrating human response with the application of daylighting systems and shading and electric light control strategies.
- Advanced Storage Concepts for Solar Thermal Systems in Low Energy Buildings: The main objective of this Task is to contribute to the development of advanced storage solutions in thermal solar systems for buildings that lead to high solar fraction up to 100% in a typical 45N latitude climate.
- Solar Heat for Industrial Process: The objective of this Task is to improve conditions for the market introduction of solar heating sys-tems for industrial applications in order to promote a reduction of fossil energy consumption and thereby to develop an environmentally friendly way of industrial production
- Testing and Validation of Building Energy Analysis Tools: this task will investigate the availability and accuracy of building energy analysis tools and engineering models to evaluate the performance of solar and low-energy buildings. The scope of the Task is limited to building energy simulation tools, including emerging modular type tools, and to widely used solar and low-energy design concepts. Activities include development of analytical, comparative and empirical methods for evaluating, diagnosing, and correcting errors in building energy simulation software.
- PV/Thermal Solar Systems: The objectives of this Task is to catalyse the development and market introduction of high quality and commercial competitive PV/Thermal Solar Systems and to increase general understanding and contribute to internationally accepted standards on performance, testing, monitoring and commercial characteristics of PV/Thermal Solar Systems in the building sector.

5.7 Solar Power and Chemical Energy Systems[12]

Concentrated Solar Power (CPS) technologies use large, sun-tracking mirrors to concentrate solar radiation. However, the final steps of generating electricity using CSP systems is similar to conventional electricity generation - the ultimate energy conversion process depends on the use of steam or gas to rotate turbines, or move a piston in a Stirling engine. In a CSP system, however, steam or hot gas is produced by the concentrated solar radiation. CSP technologies have been constructed in various sizes, from small multi-kW systems, to large power stations of several MW. These power stations have provided the cheapest electricity to be generated using solar power. The IA has three active tasks and fourteen participating countries plus the European Commission: Algeria, Australia, Brazil, Egypt, France, Germany, Israel, Mexico, Russia, South Africa, Spain, Switzerland, the U.K. and the USA. The task on Concentrating Solar Electric Power Systems deals with the design, testing, demonstration, evaluation and application of solar thermal electric systems, including parabolic troughs, power towers and dish/engine system. The task on Solar Chemistry Research is devote to the development of engineering aspects of pre commercial and demonstrational solar chemical systems projects, and basic research on solar-specific chemical reactions and processes. Finally, the annex on Concentrating Solar technology and applications is focused to develop/test solar components and subsystems; to refine computation/measurement techniques and facilities; to advance specific solar technology areas.

5.8 Wind energy[13]

The mission of the Wind Energy Systems is to stimulate co-operation on wind energy research and development and to provide high quality information and analysis to member governments and commercial sector leaders: addressing technology development and deployment and its benefits, markets and policy instruments.

The Agreement has five active tasks plus one in preparation. Nineteen countries plus the European Commission have joined it: Australia, Austria, Canada, Denmark, Finland, Germany, Greece, Ireland, Italy,

[12] www.solarpaces.org

[13] IEA wind website: www.ieawind.org

Japan, Mexico, Netherlands, New Zealand, Norway, Spain, Sweden, Switzerland, the United Kingdom and the USA.

The Agreement has a purpose to produce objective information and analysis that will inform government policy rather than directly generating policy advice. The Wind Energy Systems Implementing Agreement is expanding both internal and external information exchange. It publishes newsletters presenting results from task work, joint actions, recommended practices, analysis of implementation progress and policies are published and described in some public conferences and forums. The Wind Energy Implementing Agreement enables highly informed exchanges on national government supported programmes and findings, and is ideally placed to establish effective collaboration on basic research. As well as government sponsored R&D, considerable effort and resources are spent within the commercial sector through manufacturing industry, developers, consultancy services and in providing the physical infrastructure. The activities of the implementing agreement provides a means for international co-operation that can only accelerate cost reduction and enable more rapid deployment.

The on going and recently completed annexes deal with the Base technology information exchange; the annual review of progress in the implementation of wind energy; wind turbine round robin test program; database on wind characteristics, enhanced field rotor aerodynamics database; wind energy in cold climates; horizontal axis wind turbine aerodynamics and models from wind tunnel measurements; dynamic models of wind farms for power systems.

6. CONCLUSIONS

- The International Energy Agency, created in 1974 under the OECD umbrella, has a program for international energy R&D co-operation, that includes, inter alia, a legal mechanism called Implementing Agreement.
- The seven-article Framework provides the legal commitments of the participants and the management structure to guide the work.
- Participants to the IAs fall into two categories: Contracting Parties and Sponsors.
- The financial arrangements for international co-operation through the IAs fall into two broad categories: cost sharing, in which participants contribute to a common fund to finance the work and

task sharing, in which participants assign specific resources and personnel to carrying out their share of the work.

- The benefits of the International Energy Technology Co-operation are:
  - Shared costs and pooled technical resources;
  - Avoided duplication of effort and repetition of errors;
  - Harmonized technical standards;
  - A network of researchers;
  - Stronger national R&D capabilities;
  - Accelerated technology development and deployment;
  - Better dissemination of information;
  - Easier technical consensus;
  - Boosted trade and exports.

**** Since February 2005 in ENEA, Italian National Agency for New technologies, Energy and the Environment, 76 Lungotevere Thaon di Revel, Rome, Italy***

Chapter 5

MARKET DEPLOYMENT OF RENEWABLE ENERGY IN CENTRAL ASIA: IMPLICATIONS FOR ENERGY DIVERSIFICATION

Teresa Malyshev, PhD

Renewable Energy Unit, International Energy Agency
Paris, France
teresa.malyshev@iea.org

The diversification of energy supply toward renewable energy is economically attractive in Central Asia. Exploiting the cheapest technologies would not add to costs but would save money. IEA countries will pay for the research and development for high cost renewables. As technologies develop and costs decline, other countries can cost-effectively deploy them. Moreover, policies that have proven effective in IEA countries can be implemented in Central Asian countries. The renewable energy technologies that could be deployed on a larger scale include: biomass heating; biomass electricity; solar thermal heating; solar hot water and hydropower.

From 1990 to 2001, total primary energy supply in IEA countries grew by 1.6% per year, while renewable energy supply grew by 2.2% per year (see *Renewable Energy: Market and Policy Trends in IEA Countries* at www.iea.org). Most of the growth in renewables occurred in the 1970s and 1980s, as a result of R&D and policy support following the oil price crises. Renewable energy grew from 142 Mtoe in 1970 to 281 Mtoe in 2001, but its share in total primary energy supply shrank from 6% to 5.5% over the same period. The share in total electricity generation fell even faster, from 24.1% in 1970 to 15.1% in 2001. From 1990 to 2001,

A. Iacomelli (ed.), Renewable Energies for Central Asia Countries: Economic, Environmental and Social Impacts, 59–73.

hydropower and geothermal exhibited limited growth, and biomass grew at a rate one half of previous decades. Wind and solar, however, grew by 23% per year. These growth rates mask trends among IEA countries, Austria, Sweden and Denmark experienced rapid growth of renewables.

Policy instruments can be categorised into four quadrants, based on the direction of their support. Policies can be directed towards consumers (demand-side) or producers (supply-side). They can also be directed towards capacity (*i.e.,* the facility and/or its capital costs) or generation (*i.e.*, the product and/or the associated price to the customer). In some cases, the same policy can appear in more than one quadrant. In addition to these policy instruments, there are administrative or regulatory procedures that are not financial in nature, but nevertheless contribute to the market deployment of renewables. There are also public awareness programmes that support market deployment.

Renewable energy policies in IEA countries exhibited a clear evolution over the past three decades. Solar and wind technologies, as well as modern forms of bioenergy, began to be introduced into the market in the late 1970s after a period of research, development and demonstration that started in 1970 and intensified as the impacts of the oil price crises mounted. Denmark, Finland and Germany were the first countries to allocate RD&D funding for renewable technologies. The United States, the United Kingdom and Japan also funded renewable energy RD&D programs in the early 1970s. Except for a few countries, RD&D spending on renewable energy represented the first tier of support to the development of renewable energy markets. In the late 1970s and early 1980s, the United States, Denmark, Spain, France and Italy instituted the first market deployment policies in IEA countries.

The 1980s to the mid-1990s represented a period of experimentation with market deployment strategies among IEA countries, however, lower fossil prices reduced the urgency of the effort. Some countries employed a wide range of policies, including new variations of guaranteed prices such as feed-in tariffs combined with investment incentives and tax measures. The late 1990s saw a number of countries institute obligations and quota-based renewable energy policies as they strived to open and widen renewable energy markets. The latest development of renewable energy deployment policies is the introduction of tradable certificates.

While a clear evolution of overall policies and measures in IEA countries is evident from the figure, it does not necessarily denote a single trend. Currently, there are two main schools of thought on market deployment policies – guaranteed prices and obligations with certificates. In establishing these strategies, policymakers are working to balance two competing ideas – providing stable, attractive rates of return to attract private sector investment, and maintaining flexibility to ensure that costs are minimised and fall over time.

To identify and exploit real market opportunities, it is necessary to assess the competitiveness of specific applications and services in specific local circumstances. It is only by taking advantage of these unique situations - be it large-scale on-grid applications or niche markets off-grid, or in specific country situations - that renewable electricity can build a vigorous and sustained market. (see *Renewables for Power Generation: Status and Prospects* at www.iea.org). Many best cases already show that under optimal conditions - *i.e.* optimised system design, siting and resource availability - electricity from biomass, small hydropower, wind and geothermal power plants can be produced at low costs ranging from 2 to 5 USD cents per kWh. Cost competitiveness is then at its best, and renewable power - even without adding environmental or other values that could be attributed to certain kinds of renewable electricity generation - can compete on the wholesale electricity market.

Renewable energy projects are often the most economical choice because of their scale. Their smaller and modular nature means they can be deployed and paid for as energy demand grows; and they can be embedded within the existing energy supply networks, if they exist.

More needs to be done to diversify the global renewables industry. Energy demand is set to grow over the long-term, but do governments want fossil fuels to drive economic growth? In Central Asia, renewable energy can play a role by: freeing up natural gas for export; by improving the environment; by increasing energy efficiency through the modernising and upgrading of obsolete production capacities; by creating employment; and by enhancing energy security.

Policy makers in Central Asia could introduce specific measures to enhance the development of renewable energy (see *Renewables in Russia* at www.iea.org). This will not necessarily require substantial financial support because there are practical low-cost and often competitive measures that would stimulate investment in renewable energy technologies and could lead to considerable economic returns. In the short term, policy makers could concentrate on measures that would enhance the use of renewable energy systems that already have competitive advantages in specific applications. As businesses become experienced with installation and maintenance on a large scale, new markets for these technologies will open up, creating even more competitive opportunities.

Growth in demand for renewable energy is highest in countries that have made renewable energy development one of the focal points of the national energy strategy. To attract private investment for the implementation of this sub-programme (and more generally of other renewable energy projects), it will be necessary to put in place a clear and comprehensive legal and regulatory framework as well as specific policies and measures to stimulate the renewable energy market diffusion. Many countries have adopted specific laws that provide the legal basis for the mechanisms that allow market players to develop a renewable energy market. The renewable energy law or its equivalent usually specifies the legal status of the producers of renewable energy technologies, their rights and obligations. It also specifies the roles and responsibilities of the federal, regional and local bodies regarding such functions as establishing regulations, standards, licensing, taxation and other controls on project development.

The next step in the renewable energy strategy would be bringing into practice the mechanisms on the territorial, regional and local levels. This will require adequate regulatory and institutional frameworks to actually set the rules authorised under the national renewable energy policy. A set of regional laws or regulations and local provisions are necessary to guarantee the implementation of the national strategy.

Since renewable energy technologies are very site-specific, the success of a project will depend on the exact resource conditions at the specific site. For this reason preliminary local site monitoring is very important to stimulate the development of renewable energy projects. Regional atlases of renewable energy resources should be published and disseminated for potentially interested parties, such as energy companies, electricity utilities, local industries and authorities, and the general public.

Many countries have succeeded in building public awareness through information campaigns to inform professionals and the general public about the existing renewable energy resources, technologies, applications and their environmental and public benefits. International experience demonstrates that increasing public awareness of renewable energy can lead to a significant increase in the use of renewable energy in the residential sector, especially in the case of biomass and solar thermal. Different means of disseminating information should be used: the mass media, organisation of conferences and publication of outlets. Success stories should be widely presented to the mass media. Joint ventures are also an effective way of sharing technological expertise. Joint ventures also allow companies to improve their managerial, financial and commercial skills, while providing foreign companies with a highly-skilled, low cost partner. Russia partnered with Denmark and with US for wind installations.

It is important to efficiently design financial programs, including: reduced or zero VAT; investment tax incentives; accelerated depreciation; favourable loans; demonstration projects; and environmental funds.

Regulatory policy also has to be tailored to increase the competitiveness of renewable energy. For example, Renewable Portfolio Standards oblige power companies to supply or customers to buy a certain percentage of electricity from renewable energy sources. Frequently, this mechanism is combined with a national renewable energy target. For example, in 2000, Poland adopted a regulation which requires all power and heat companies to buy energy from renewable sources. To fulfill this obligation, each power company had to provide at least 2.4% of total electricity sales from RE sources in 2001.

Expanding grid-connected bulk power generation from renewables depends to a large extent on the conditions and rates under which independent power producers can gain access to the transmission system Specific regulations and implementation frameworks are necessary to allow IPPs to operate effectively in the market. The status and rights of independent power producers have to be clearly defined. Industries which produce their own electricity must have non-discriminatory access to the grid to transmit the generated electricity for their own use or for sale.

Net metering is a policy for customers with small, grid-connected renewable generating facilities. It allows the electric meters of these customers to operate in reverse when the systems produce energy in excess of the customer's demand. This enables customers to use this generation to offset their consumption over a longer period of time. Through the offset, customers receive the full retail value for the excess electricity they generate, similar to the value customers receive when they conserve electricity.

Wholesale market competition, for which lowest-cost power is generally the primary driver, may reduce incentives to develop renewable energy because of the higher capital costs of many technologies. Also, the inability to dispatch power on demand with some renewables is a drawback in spot markets, in which a high premium is placed on generators that can assure power availability during peak periods. Different countries have adopted special regulations to guarantee that electricity and heat services are provided in rural areas, where the cost of these services is generally higher than in urban areas.

Diversifying energy supply with renewables can help to achieve the three E's (economic growth, environmental improvement and energy security) in Central Asia. Policies in IEA countries are driving the growth in renewable energy markets worldwide. And, today, there are many cost-effective opportunities to exploit renewables in Central Asia.

Figure 1.

MAIN POINTS

- **Why does diversifying energy supply with RE help achieve 3Es?**
- **What's happening in RE markets?What policies are driving growth?What are opportunities for RE in Central Asia?What are strategies for market deployment of renewables?**

Figure 2.

WHY DIVERSIFY TOWARDS RENEWABLES?

- **Cheapest energy source**
- **In certain locations**
- **For certain technologies**
- **Creates local jobs**
- **Improves environment**
- **Export potential**
- **Oil & gas revenue**
- **RE exports**

Figure 3. IEA Total Primary Energy Supply 1970-2001

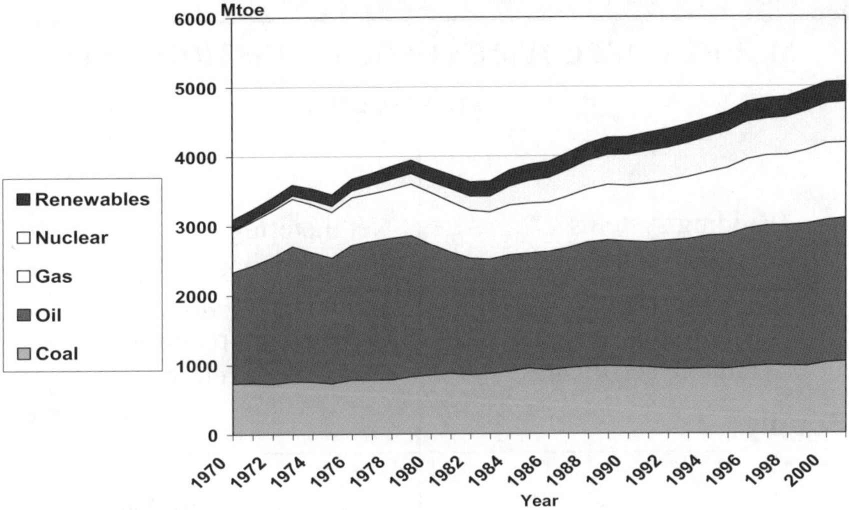

Figure. 4

How Did Renewables Perform?

Good news:

- **142 Mtoe in 1970 to 281 Mtoe 2001**

Bad news:

- **share in TPES shrank from 6 % in 1992 to 5.5 % in 2001**
- **share in electricity production fell from 24.1% in 1970 to 15.1% in 2001**

Figure. 5

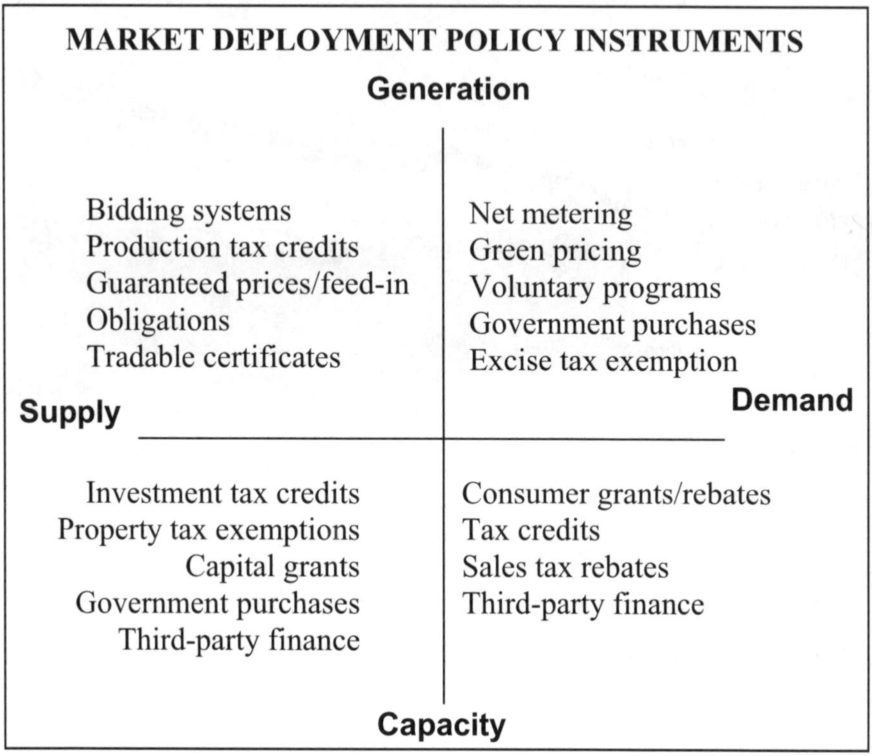
MARKET DEPLOYMENT POLICY INSTRUMENTS
Generation
Bidding systems
Production tax credits
Guaranteed prices/feed-in
Obligations
Tradable certificates
Net metering
Green pricing
Voluntary programs
Government purchases
Excise tax exemption
Supply
Demand
Investment tax credits
Property tax exemptions
Capital grants
Government purchases
Third-party finance
Consumer grants/rebates
Tax credits
Sales tax rebates
Third-party finance
Capacity

Figure. 6

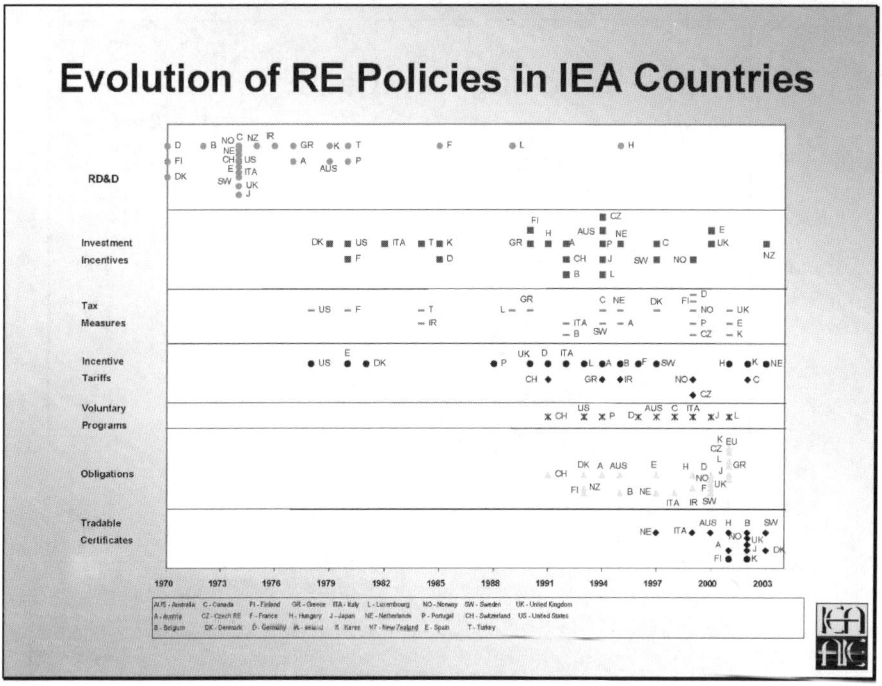

Figure. 7

Figure. 8

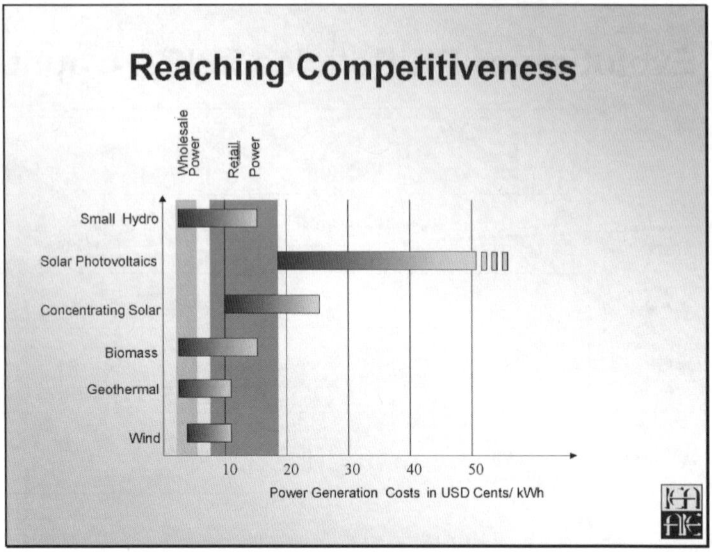

Figure. 9

HOW DID RENEWABLES PERFORM?

Good news:

- **almost all IEA countries have enacted support schemes by end 90s.**

Bad news:

- **85 % of wind energy capacity concentrated in 4 countries (Germany, Denmark, Spain, USA);**
- **86 % of PV concentrated in 3 countries (Germany, Japan, USA).**

Figure. 10

ROLE FOR RENEWABLES IN CENTRAL ASIA

Challenges:

- **High dependence on fossil fuels**
- **Overcapacity in electricity production**
- **Inefficient energy use**
- **Low prices for conventional energy**
- **Lack of commercial know-how**

Figure. 11

ROLE FOR RENEWABLES IN CENTRAL ASIA

Opportunities & Benefits:

- **Diversification – regional dimension**
- **Diversification - economics**
- **Environmental implications**
- **Employment effects**
- **Aging electricity infrastructure**
- **Lower cost production of RES**

Figure. 12

HIGHLIGHTING THE ROLE OF RENEWABLES: 4 KEY ISSUES

- ♦ **Frees up oil and gas for export**
- ♦ **Substitutes for fossil fuels**
- ♦ **Reduces cost of supplying energy to some users**
- ♦ **Increases local employment**

Figure. 13

(1) INCREASES GAS EXPORT POTENTIAL

- ♦ **Uptake of renewables can free up gas for export**
- ♦ **EU gas imports expected to rise from 233 billion cm in 2002 to 639 billion cm in 2030**
- ♦ **Import share (EU) – 81% in 2030**

Figure. 14

(2) SUBSTITUTES FOR FOSSIL FUELS

- ♦ **Biofuel District Heating in Lithuania**
- ♦ **Heat production costs down nearly 20%**
- ♦ **Local production of biofuel saved $624,000 in avoided fuel imports**
- ♦ **Reduction in CO_2, SO_2, NO_x emissions**

Figure. 15

(3) REDUCES COST OF SUPPLYING ENERGY

- **Off-grid solution**
- **Replaces diesel generators**
- **Avoids transport difficulties**
- **Heating and hot water solution**
- **Replaces conventional fuels**
- **Creates local industry**

Figure. 16

(4) CREATES LOCAL EMPLOYMENT					
Wind	Geothermal	Solar PV	Solar Thermal	Landfill Gas	Natural Gas (PG)
Construction Jobs					
2.57	4.00	7.14	5.71	3.71	1.02
Operation and Maintenance Jobs					
0.29	1.67	0.12	0.22	2.28	0.13

Figure. 17

BUILDING RE MARKETSDeveloping national strategy to encourage cost-effective solutions

- **Stimulating demand and supporting domestic industry**
- **Designing financing programs around specific technologies**
- **Tailoring regulatory policies**

Figure. 18

DEVELOPING A NATIONAL STRATEGY

- **National Renewable Energy Strategy**
- **National legal framework**
- **Regional laws and regulations**

Figure. 19

STIMULATING DEMAND

- **Analysis of RE resources**
- **Public Awareness of ALL benefits**
- **International technology partnerships**

Figure. 20

DESIGNING FINANCIAL PROGRAMS

- **Reduced or zero VAT**
- **Investment tax incentives**
- **Accelerated depreciation**
- **Favourable loans**
- **Demonstration projects**
- **Environmental funds**

Figure. 21

TAILORING REGULATORY POLICY

- **Portfolio standards**
- **Grid access**
- **Net metering**
- **Power purchase agreements**
- **Energy supply in rural areas**

Figure. 22

CONCLUSION

- **lDiversifying towards renewables can help achieve 3Es**
- **Policies are driving growth in RE markets worldwide**
- **There are cost-effective opportunities to exploit renewables in Central Asia**

Chapter 6

WORLD ENERGY OUTLOOK 2004: KEY FINDINGS AND MESSAGES

Marco Baroni

International Energy Agency

Soaring oil and gas prices, the increasing vulnerability of energy supply routes, massive investment requirements and increasing emissions of climate-destabilising carbon dioxide keep characterising the world energy context.

The Reference Scenario of the IEA's World Energy Outlook 2004 depicts the future trends, from nowadays to 2030, of global energy demand in the absence of new energy policies after those legally enacted by mid 2004. In this scenario, world energy requirements in 2030 will reach 16.5 billion tonnes of oil equivalent, almost 60% higher than in 2002. The average annual growth rate of 1.7% will be slower than the growth rate of the last 3 decades of 2%.

In 2030, fossil fuels will continue to dominate the global energy mix, accounting for 85% of the increase of the world energy demand (see figure 1).

Oil will continue to be the largest fuel among the energy sources, with a share slightly decreasing from 36% in 2002 to 35% in 2030. In this year, the transport sector will require 54% of the 5766 million tonnes of oil equivalent (Mtoe), accounting for about two-thirds of the global increase of oil use.

Natural gas demand growth will be the strongest among all fossil fuels, at an average of 2.3% per year over the projection period. The share of natural gas in the total energy needs will increase from 21% in 2002 to 25% in 2030, overtaking coal as second largest energy source. The power sector will be the main driver of gas consumption, accounting for 60% of the

A. Iacomelli (ed.), Renewable Energies for Central Asia Countries: Economic, Environmental and Social Impacts, 75–85.

increase from 2002 to 2030, with its share rising from 36% in 2002 to 47% in 2030 of the world gas market.

Figure 1: World Primary Energy Demand by Fuel

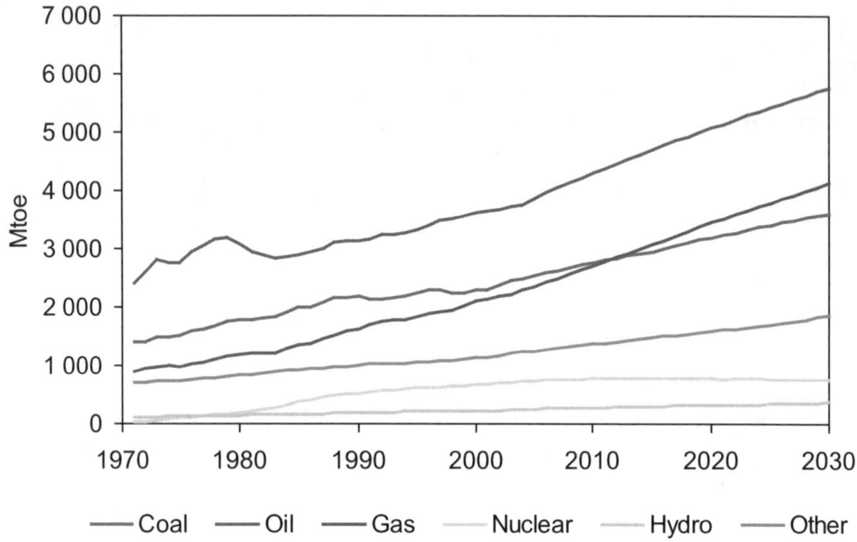

Coal will be the slowest-growing among the fossil fuels, at 1.5% per year, although continuing to play a key role in the world energy mix. Its share will decline slightly from 23% in 2002 to 22% in 2030. The power sector will be responsible for almost four-fifths of the global coal demand in 2030, accounting for almost all of the increase over the next 3 decades. China and India will account for the majority of this increase over the period 2002 to 2030, while demand growth in the OECD will be minimal.

Nuclear energy will remain substantially stable over the projection period, with its share falling from 7% in 2002 to 5% in 2030. The rate of construction of new reactors is expected to keep pace with the rate at which old reactors will be retired, with the majority of retirements in the OECD countries and the bulk of the new additions taking place in Asia.

Hydropower production will expand by 1.8% per year over the projection period, a slightly faster rate than that of global primary energy demand, therefore maintaining constant, at about 2%, its share in the total energy needs.

The role of biomass and waste, the use of which is concentrated mainly in the developing countries, will gradually diminish over the projection period. Globally, its share of primary energy demand will decrease from 11% in 2002 to 10% in 2030, as it will be replaced with modern fuels. In absolute terms, though, the consumption of traditional biomass in developing countries will continue to grow over the projection period.

The use of the non-hydro renewables (excluding biomass) will increase fastest among all energy sources, at an annual growth rate of 5.7%, although its share in 2030 will still be small, at only 2%.

Renewable energy sources as a whole will increase their share of electricity generation throughout the projection period. Although hydro power generation will increase by over 60%, its share will fall, but the shares of other renewables in electricity generation will triple, from 2% in 2002 to 6% in 2030. While hydropower will remain the largest source of electricity generation by renewables in 2030, the contribution of wind and biomass will also become quite substantial (see figure 2).

Figure 2: World Electricity Generation from Non-Hydro Renewable Energy Sources

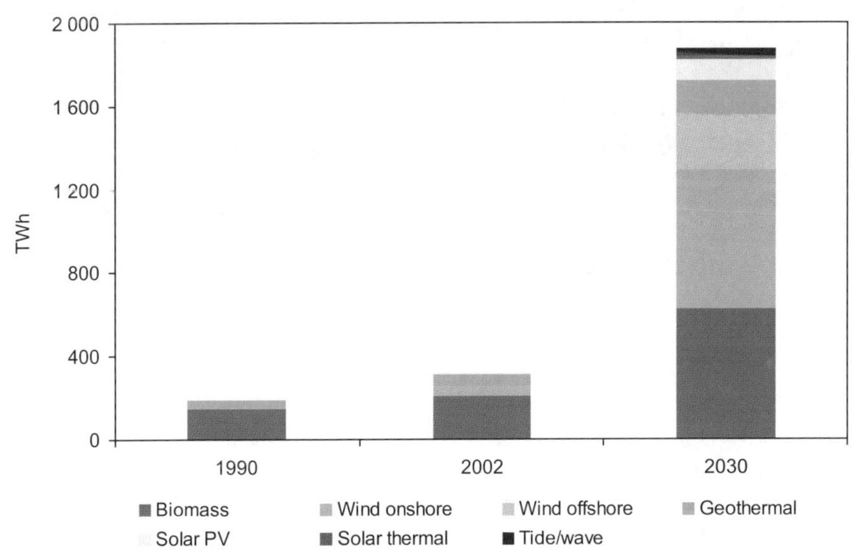

Two-thirds of the increase in global energy demand between 2002 and 2030 will come from developing countries. By 2030, they will account for almost half of total demand, in line with their more rapid economic and population growth. OECD countries will account for 26% and the transition economies for the remaining 8% of the demand growth. Consequently, the current 52% share of the OECD in world demand will decline to 43% in 2030, while that of the developing countries will increase, from 37% to 48% (see figure 3). The developing countries' share of global demand will increase for all of the primary energy sources except non-hydro renewables. Their share of nuclear-power production will increase fastest, because of strong growth in China and other parts of Asia. Their share of coal consumption will also increase sharply, mainly because of booming demand in China and India.

Figure 3: Regional Shares in World Primary Energy Demand

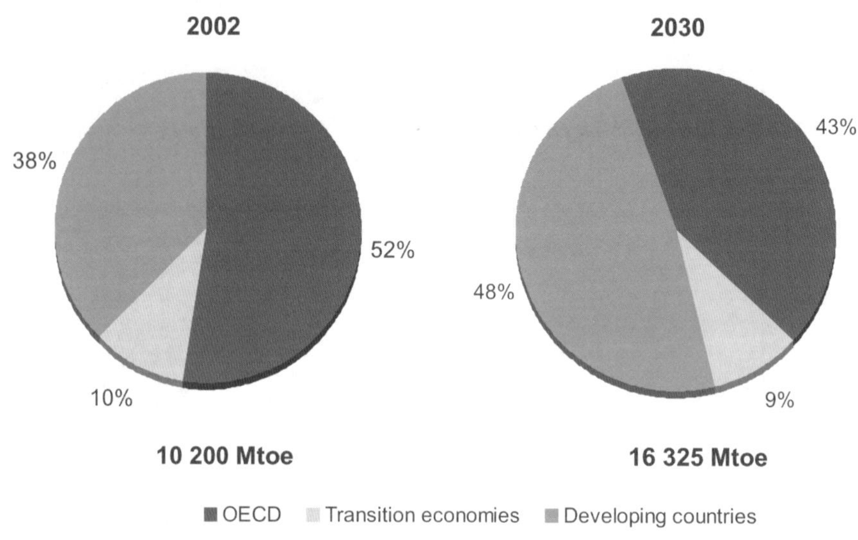

The world's energy resources are adequate to meet the projected increase in energy demand until 2030 but serious concerns about energy security emerge from the market trends projected here, as energy production will be more and more displaced with respect to energy consumption. The geographical sources of incremental energy supplies will shift markedly over the projection period, mainly in response to cost factors and the location of resources. From 2002 to 2030, more than 95% of the increase in production will occur in non-OECD regions, against about 70% from 1971 to 2002. Most low-cost fossil-fuel resources are located in non-OECD countries. This situation will more and more increase energy trade and will raise concerns about security of supply and vulnerability to supply disruptions. Energy exports from non-OECD to OECD countries will increase by more than 80%, from some 1 500 Mtoe in 2002 to over 2 700 Mtoe in 2030. Trade between countries within each grouping is also expected to grow. As international trade expands, risks will grow of a supply disruption at the critical chokepoints through which oil must flow. Traffic through the Straits of Hormuz and Malacca, where currently a total of 26 million barrels pass through every day, is projected to more than double over the projection period. A disruption in supply at any of these points could have a severe impact on oil markets.

If current government policies do not change, energy-related emissions of carbon dioxide will grow marginally faster than energy use. CO_2 emissions will be more than 60% higher in 2030 than now. Well over two-thirds of the projected increase in emissions will come from developing countries, which will remain big users of coal – the most carbon-intensive of fuels. Power stations, cars and trucks will be responsible for most of the increased CO_2 energy-related emissions, which will increase by 1.7 % per year over 2002-2030, reaching 38 billion tonnes in 2030, an increase of 15 billion tonnes with respect to the 2002 levels.

Over the past three decades, energy-related CO_2 emissions worldwide have grown less rapidly than has primary energy demand, as carbon emissions grew by 1.7% per year, while energy demand grew by 2%. Over the projection period, emissions and demand will grow at about the same rate, 1.7% per year, (Figure 4). The average carbon content of primary energy consumption will remain more-or-less constant at about 2.3 tonnes of CO_2 throughout the projection period.

Figure 4: Average Annual Growth in World Primary Energy Demand and Energy-Related CO2 Emissions

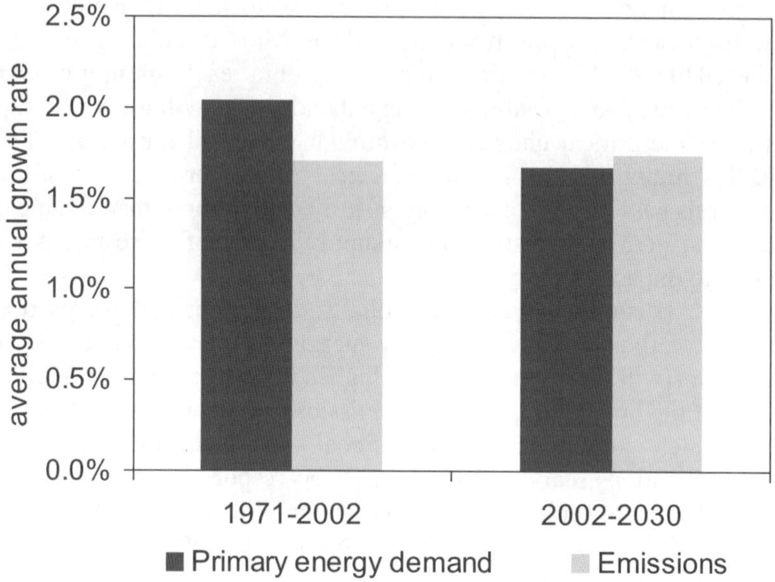

Developing countries will be responsible for about 70% of the increase in global CO2 emissions from 2002 to 2030 (Figure 5). They will overtake the OECD as the leading contributor to global emissions early in the 2020s. OECD countries accounted for 54% of total emissions in 2002, developing countries 36% and transition economies for 10%. By 2030, the developing countries will account for 49%, the OECD countries for 42% and the transition economies for 9%. Today, developing-country emissions are two-thirds of OECD emissions. By 2030, they will be 16% higher.

Figure 5: World Energy-Related CO2 Emissions by Region

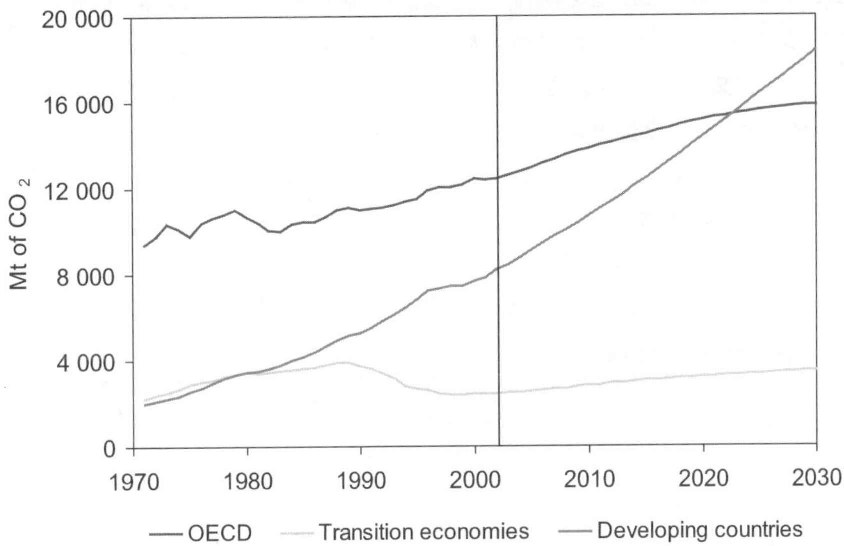

Converting the world's resources into available supplies will require massive investments. In some cases, financing for new infrastructure will be hard to come by. Meeting projected demand will entail cumulative investment of some \$16 trillion (in year-2000 dollars) from 2003 to 2030, or \$568 billion per year. The electricity sector will absorb most of the future energy investments: power generation, transmission and distribution will absorb almost \$10 trillion, or 62%, of total energy investment (see figure 6). Of this, more than half of the investment in the electricity industry will go to transmission and distribution networks. If we include the investment needed in the fuel chain to meet the fuel needs of power stations, electricity's share rises to more than 70%. Total investments in the oil and gas sectors will each amount to almost \$3 trillion, or around 18% of global energy investment. Exploration and development will take more than 70% of total investment in oil, while the share is lower for gas (at 56%) because transportation infrastructure needs for gas are bigger. Coal investment will amount to only \$400 billion, or 2.5% of the total. Coal is about a sixth as capital-intensive as gas in producing and transporting a given amount of energy. Developing countries, where production and demand are set to increase most, will require about half of global energy investment. Those countries will face the biggest challenge in raising finance, because their needs are larger relative to the size of their economies and because the investment risks are bigger.

Figure 6: Cumulative Energy Investment, 2003-2030

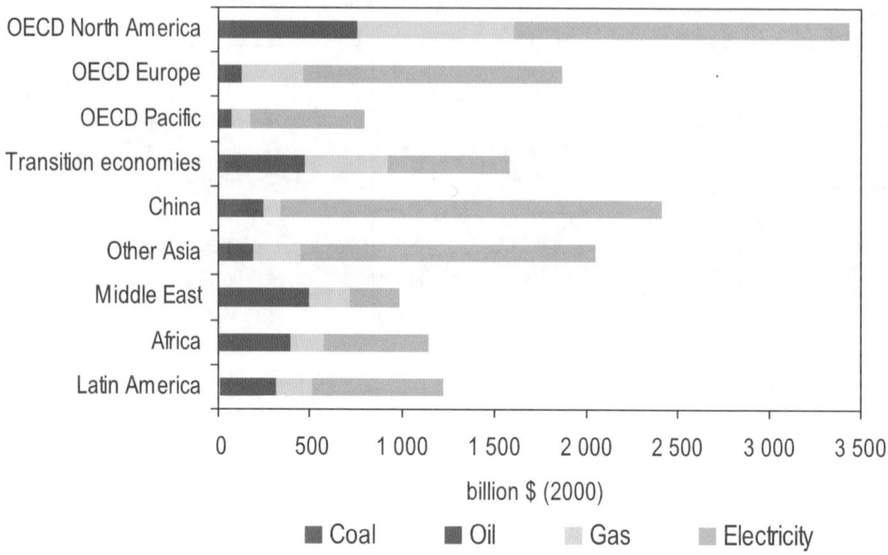

The World Energy Outlook 2004 presents also a World Alternative Policy Scenario, which depicts a more efficient and more environment-friendly energy future than does the Reference Scenario. It analyses how global energy trends could evolve if countries around the world were to implement a set of policies and measures that they are currently considering or might reasonably be expected to adopt. These policies would promote the faster deployment of more efficient and cleaner technologies, reducing to 14.7 Gtoe the global energy demand in 2030, about 10% lower than in the Reference Scenario (see figure 7). Energy demand in this scenario grows by 1.3% per year, 0.4 percentage points less than in the Reference Scenario.

The fuel mix is markedly different, with a reduction of 14% of fossil fuels, an increase of nuclear power (+14%) and a substantial expansion of non-hydro renewable energy sources (+30%). By 2030, carbon-free fuels account for 22% of global primary energy demand, four percentage points higher than in the Reference Scenario.

Coal consumption sees the biggest reduction with respect to the Reference Scenario, with a drop of almost a quarter, or 857 Mtoe, mostly driven by lower demand from the power sector. This decrease would be primarily due to increased efficiency of power plants – especially in developing countries – and fuel-switching to less polluting energy sources.

The average annual rate of growth in coal demand would be 0.5%, down from 1.5% in the Reference Scenario.

In the Alternative Scenario oil demand increases to just less than 5 Gtoe in 2030, or 11%, lower than in the Reference Scenario. The reduction of 12.8 mb/d is an amount equal to the current combined production of Saudi Arabia, the United Arab Emirates and Nigeria. The transport sector would account for almost two-thirds of the savings, thanks to increased fuel efficiency and faster penetration of alternative-fuel vehicles that would push demand down.

Natural gas demand is 10% lower in 2030 than in the Reference Scenario. Again, the power sector accounts for most of the savings, almost three-quarters by 2030. Consumption of traditional biomass decreases, thanks to more efficient use in industrial processes and in household cook-stoves, but it is offset by the increased consumption in the power sector and by biofuels use in the transport sector, thus raising total biomass consumption of 43 Mtoe with respect to the Reference Scenario. Consumption of other renewables increases even more, adding 75 Mtoe in 2030 – a 30% increase compared to the Reference Scenario. Power generation drives most of this increase, but solar water heaters and geothermal energy also contribute.

Figure 7: Energy Demand in the Reference and Alternative Scenarios

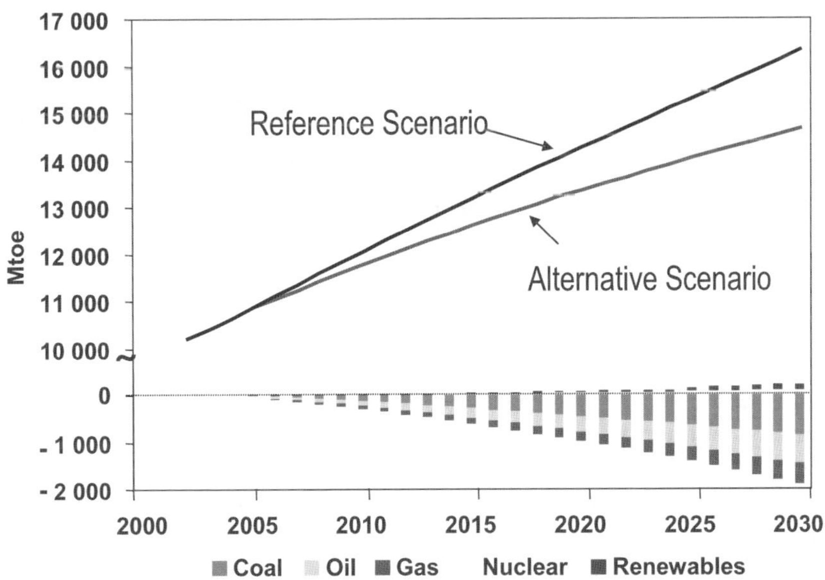

As a result of the lower level of energy demand and of the different fuel mix, energy-related CO2 emissions in the Alternative Policy Scenario reach 31.7 billion tonnes in 2030, about 6 Gt, or 16%, lower than in the Reference Scenario (Figure 8). They still increase at a global level by 37% from 2002 to 2030, but at a growth rate of 1.1% per annum with respect to the 1.7% of the Reference Scenario. The gap is particularly wide in the third decade, when the annual growth rate is halved, from 1.4% to 0.7%. The reduction is comparable to the current combined emissions of the United States and Canada.

Figure 8: Global Energy-Related CO2 Emissions in the Reference and Alternative Scenarios

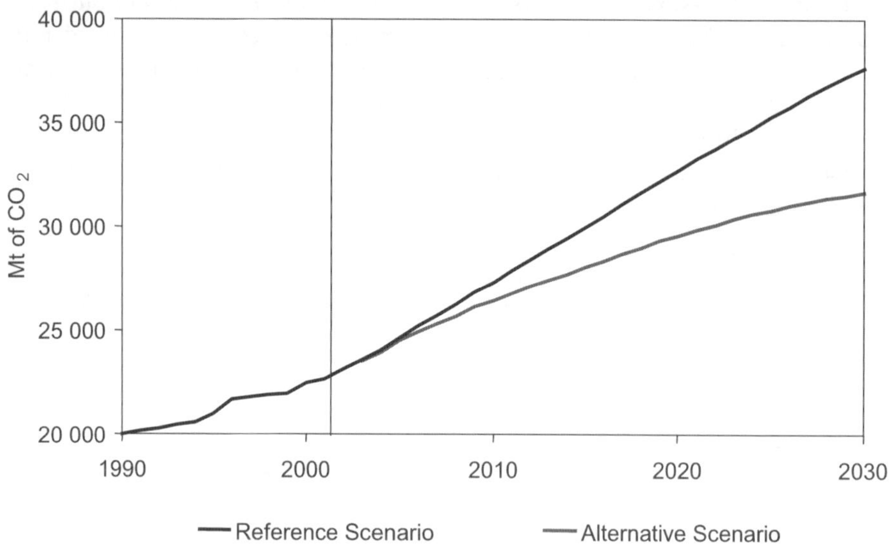

The difference between the growth rates of CO2 emissions in the two scenarios is summarised in Figure 9. Measures to improve end-use efficiency explain almost 60% of the difference worldwide. These measures include more efficient vehicles, industrial processes and appliances, as well as stricter building standards. In the transition economies and in the developing countries, the role played by energy-efficiency measures is particularly large, reflecting the huge potential for efficiency improvements there. The other big contributor to lower emissions is the increased share of renewables in power generation, accounting for 20% of the global reduction. The increased role of nuclear power accounts for an additional 10% and fuel switching in end-uses and switching from coal to natural gas in power generation explain the rest.

Figure 9: Reduction in Energy-Related CO2 Emissions in the Alternative Scenario by Contributory Factor, 2002-2030

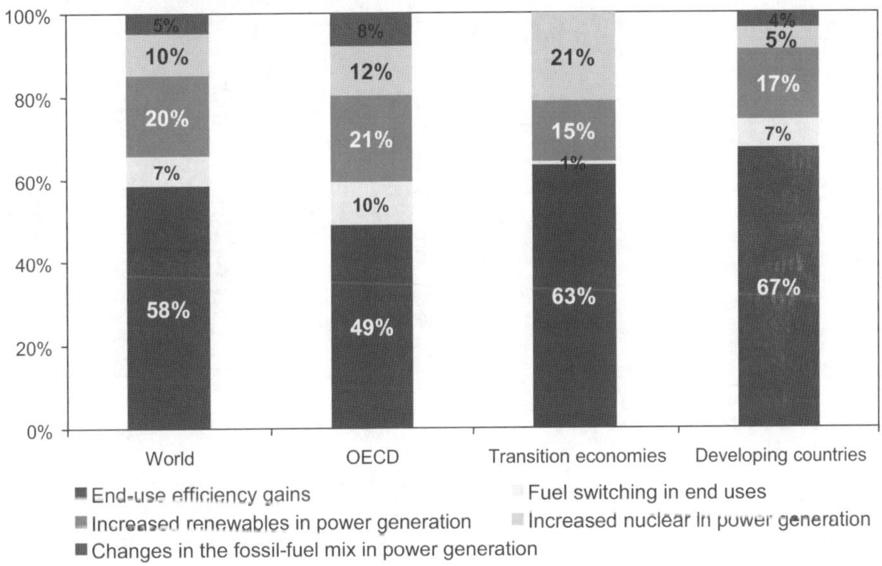

Chapter 7

MAIN ACHIEVEMENTS OF THE IEA PROGRAMME ON HYDROPOWER TECHNOLOGIES

Frans H. Koch

former Secretary
IEA Implementing Agreement for Hydropower Technologies and Programmes

1 INTRODUCTION

The International Energy Agency has set up more than forty "Implementing Agreements" dealing with a wide variety of energy technologies. One of these is the Implementing Agreement for Hydropower Technologies and Programmes (Hydro IA). This chapter will describe what the Hydro IA has achieved since its inception in 1995. During this period, six task forces worked on the following subjects:

- Upgrading of hydropower plants
- Small scale hydropower
- Environmental and social aspects of hydropower
- Education and training of hydropower professionals
- Public awareness and public acceptance of hydropower
- Good practices in hydropower technology

The countries that participated in the Hydro IA were: Canada, China, Finland, Japan, Norway, Sweden, and the USA. The remainder of this paper will describe the work and the findings or conclusions of each of the six task forces mentioned above.

A. Iacomelli (ed.), Renewable Energies for Central Asia Countries: Economic, Environmental and Social Impacts, 87–94.

2 UPGRADING OF HYDROPOWER PLANTS

2.1 Benefits of hydro upgrading projects

The upgrading of hydropower plants refers to the replacement or rebuilding of turbines, generators, control equipment or hydraulic structures. It results in adding to the plants power capacity, usually expressed in Megawatts (MW), and/or to the amount of energy a plant can produce in a year, usually expressed in Megawatt hours (MWh). Many of the older hydro plants were built at a time when computers were not yet used for the design of turbines and generators, and when automatic control systems were not widely used. Modern turbines and generators may be from 2 to 5 % more efficient, in other words they produce from 2 to 5 % more electricity from the same amount of water. Upgrading projects generally cost only a small fraction of the amount for new generation facilities, since only one or two components of the total facility have to be re-built or replaced. For this reason the upgrading of existing hydro plants is the cheapest form of renewable electricity available, in some cases the cost was US $ 70 per kW of capacity. A new facility would typically cost from $ 800 to $ 2000 per kW of capacity.

There are several thousand hydro plants in the world with a capacity of more than 100 MW, and several dozen with a capacity of more than 2000 MW. Adding 2 % to the capacity of such plants through upgrading projects represents 2 MW in the first case and 40 MW in the second. Geothermal or bio-energy electricity generation plants can also be in this size range of 2 to 40 MW. However, the typical size of other renewable energy technologies tends to range from 0.01 MW to 2 MW. Compared to these, upgrading of hydro plants is both large scale and low cost.

2.2 Work done by the upgrading task force

The task force on upgrading produced three technical reports in CD-ROM format to assist engineers and project managers. They are sold through the publisher of Hydro Review Worldwide (March 2001, HCI publications, Kansas City, MO USA). The titles are:

1. Guidelines on Methodology for Hydroelectric Francis Turbine Upgrading by Runner Replacement.
2. Guidelines for Hydroelectric Generator Upgrading.
3. Guidelines on Methodology for Control Systems Rehabilitation

3 SMALL SCALE HYDROPOWER

3.1 Description of small scale hydropower

The term small scale hydropower is defined in widely different ways in different countries. In some cases it means a capacity of 500 kW or less whereas in others it may be as high as 50,000 kW. Limits of either 2 MW (2000 kW) or 5 MW are commonly used. About 20 to 30 years ago many small hydro plants were abandoned because the cost of staffing them 24 hours per day became prohibitive. The advent of automatic supervisory and control systems in the last two decades has radically improved the economics of small hydro plants; most of them now run unattended and only need to be visited periodically for maintenance. In addition some sophisticated computer software programs now make it much easier to assess potential sites for small scale hydropower and to do some of the preliminary design.

There is a widely held perception that small scale hydropower is environmentally more benign than larger scale plants. In fact, this is highly questionable if the environmental impacts per kWh of electricity generated are considered. A large scale plant of, say 500 MW, may well have considerably fewer impacts than 100 small plants of 5 MW each. However, the perception is deeply rooted and many governments are reluctant to explicitly support large scale hydropower while they actively encourage the development of small scale hydropower.

3.2 Achievements of the small scale hydro task force

In most IEA countries small hydropower is developed by small organizations, municipalities, or entrepreneurs. The task force has produced reports and published them on the Internet (www.small-hydro.com) to facilitate the work of developers. Firstly, they have produced a global atlas that shows the location and parameters of existing small hydro plants and potential sites for new plants. The Atlas contains data for some countries (Canada, China, Norway, UK, USA) whereas others remain to be done. All countries are invited to collect these data and publish them on the Internet, either on the web-site of the task force or by borrowing the software and making it available to the public in their own language.

Secondly, the task force has published a series of reports on financing of small scale hydro projects, on planning, on environmental aspects, etc. that should assist developers with the first steps of their projects.

4 TASK FORCE ON ENVIRONMENTAL EFFECTS

4.1 Positive and Negative Effects of Hydro Projects

Any major infrastructure project or any electricity generating plant has both positive and negative effects on the environment and on society. This is also true for hydropower projects. One distinguishing feature is the site specific nature of hydropower. Each project is on a particular river within a specific ecosystem, landscape, and human settlement pattern. Some are high up in the mountains, above the tree line. Others are on rivers that flow through deserts or semi-deserts. Yet others are in the middle of large cities. Because almost every hydro project is very different from every other hydro project, the environmental and social effects are also very different. For every new project, the government and the affected people have to make a decision whether or not the benefits of the projects outweigh the adverse effects. For projects that were built in the past, some have delivered outstanding benefits with very few environmental or social disadvantages, in other cases the benefits and costs were more equal, and a few projects have become controversial because their disadvantages are considered to outweigh their benefits.

4.2 Achievements of the Environmental Task Force

The aim of the environmental task force was to produce, balanced, objective assessments of the positive and negative social and environmental effects of hydropower and of the mitigation measures that can be taken to counteract the negative effects. The task force consisted of 8 organizations from 7 countries, and over a 5 year period it held 12 international meetings and workshops in 8 different countries. In all, 112 persons from 16 countries contributed to the work and to the final reports.

Case studies that illustrated specific environmental or social challenges and ways of mitigating them were collected for 46 projects. A second part of the work was a life cycle analysis comparison of different electricity generating technologies, including coal, natural gas, diesel, oil, nuclear, hydro, biomass, wind, and solar photo-voltaic. A third part of the work collected and compared the legal and regulatory framework for environmental impact assessments in several countries. A fourth part of the work specifically studied mitigation measures and their effectiveness. Reports were published for each of these parts, and then a summary report was prepared that gave an overview of all the information collected and made recommendations to policy makers and managers of hydro projects. All these reports are available on the web-site: www.ieahydro.org. The 5 main recommendations addressed:

1. The Energy Policy Framework
2. Decision making processes
3. Comparison of alternatives for planned Hydropower projects
4. The management of hydropower plants
5. Sharing benefits with local communities.

5. EDUCATION AND TRAINING

5.1 Overview of global education and training facilities

The largest share of potential hydro projects that remain to be developed are in Asia and Africa where there are relatively few education and training facilities for hydropower managers, engineers, and technicians. On the other hand, some of the hydropower education facilities in IEA countries have spare capacity because most of their hydropower potential is already developed. A sufficient number of trained professionals is necessary, of course, if the hydropower industry is to maintain itself and grow. Meeting the education and training demand requires planning and foresight.

5.2 Achievements of the education task force

The task force started by surveying the education and training facilities that are available world wide. Based on this, it published reports outlining program requirements in operation and maintenance training and planning of hydropower projects.

The task force also worked on facilitating the exchange of teaching materials between IEA countries and developing countries using the Internet. It developed software that would safeguard the intellectual property rights of those who had prepared teaching materials, such as lectures, books, articles, designs, etc., and yet allowed the sharing of such material over the Internet. To test this concept, links were established among universities in Japan, Norway, and Sweden and Brazil, Thailand, and Zambia.

6 PUBLIC AWARENESS AND ACCEPTANCE

6.1 Public knowledge about hydropower

The public becomes involved at a local level when average sized hydro plants are proposed, or their license needs to be renewed, and at a national level for very large hydro projects. In such debates, there will usually be proponents of the project who publish positive information, and opponents of the project who publish negative information about the effects on the community and the environment. This creates a need for reliable, balanced, and objective information about hydropower. As mentioned previously, hydro projects are very different one from the other. Some projects consist of a large dam that creates a reservoir and floods a certain area of land, whereas others don't. Large dams and reservoirs were the subject of a two year study conducted by the World Commission on Dams, in which non-governmental organizations, governments, and the hydropower industry participated. The report of this commission recognized that large dams were necessary, and sometimes essential, for the social and economic development of many countries. However, the decision making processes involved and the treatment of people affected by dam projects should be improved. More could also be done to mitigate environmental effects and to safeguard cultural monuments. The United Nations Environment Program (UNEP) has implemented a follow-up project to the World Commission on Dams study, it is called the "Dams and Development Project" and aims to provide governments and civil society with the information they need to make sound decisions about projects involving large dams. In summary, many organizations throughout the world are active in educating the public about the issues relating to hydropower, and especially to large dams. These efforts require the availability of reliable, balanced, and objective information about hydropower.

6.2 Achievements of the public awareness task force

The Public Awareness task force cooperated with other national and international hydropower organizations to produce and publish objective information about hydropower. They jointly produced a white paper on hydropower that is available on the web-site www.ieahydro.org, and has been translated into German, Portuguese and Japanese. The same web-site is also the home of the "Hydropower Information Network" that provides extensive information and references on all aspects of hydropower, and is addressed to a variety of target groups ranging from the interested public to specialized hydropower managers and engineers. Another activity was the publication of a special issue of "Energy Policy" that dealt entirely with hydropower.

7 GOOD PRACTICES

7.1 Need for a report on hydropower good practices

Both hydropower professionals and the interested public need better and more accessible information on good practices in hydropower. The professionals need it because they may be faced with a particular technical, environmental, or social problem in an existing or planned project. They will want to see what has been done in other countries to deal with it. The interested public needs it because opponents of hydropower put great emphasis on the relatively few hydro projects that have experienced serious environmental or social problems. In the technical literature and in the more general media these same few projects are cited again and again. To restore the balance, it would be useful to have a report where these problems were successfully solved. The environmental task force identified ten "key issues" related to hydropower as follows:

1. Resettlement and rehabilitation
2. Culturally vulnerable communities
3. Public health risks
4. Optimization of regional development benefits
5. Biological diversity
6. Modifications to hydrological regimes
7. Dams as obstacles to passage
8. Sedimentation in reservoirs
9. Modifications to water quality
10. Creation of reservoirs

A single, easily available, report that illustrates how these 10 key issues have been successfully addressed would fill a need for all those who are interested in hydropower.

7.2 Work of the task force on good practices

The Good Practices Task Force looks forward to completing its report at the end of 2004, and publishing it in 2005. It has collected case studies for the ten key issues listed above, and attempted to get good geographic coverage by obtaining case studies from Asia, North America, South America, Europe, Australia and Africa. In addition to the ten key issues, there will also be case studies that illustrate the social, economic, and environmental benefits that some hydro projects have brought to their surrounding communities and to their countries.

8 SUMMARY

Hydropower is by far the largest source of renewable electricity available today; it produces about 17 % of global electricity. This is more than 20 times the amount produced by other renewable technologies. The future development of hydropower will require its acceptance by governments and the public, and this in turn will mean that social and environmental effects will need to be carefully managed.

Hydropower is very site specific; each project is different from most others and is situated in a particular landscape, climate, ecosystem and human settlement area. This means that few general statements can be made about hydropower, either good or bad, and each project has to be judged on its own merits.

With good planning, good operation, and good mitigation measures, hydropower can remain the principal source of renewable electricity for many decades to come. It can make a valuable contribution to our societies and to the environment.

Chapter 8

RENEWABLE RESOURCES TO HYDROGEN: APPROPRIATED TECHNOLOGIES FOR DEVELOPING COUNTRIES

V. Naso, E. Bocci, F. Orecchini, D. Marcelo

CIRPS, Interuniversity Research Center for Sustainable Development[1]
Piazza del Colosseo 9, 00184, Roma, Italy

1. RESEARCH FIELDS OF ENERGY AND ENVIRONMENT GROUP ABOUT HYDROGEN PRODUCTION FROM RENEWABLE RESOURCES

It's completely different having a certain efficiency, when the resource is renewable or non renewable on Earth.

The goal is not only the increase of efficiencies but primarily the indication of feasible cycles that, starting from a renewable resource, are able to completely "close", with no resource consumption and no impact on the environment. Renewable resources adoption as replacement of fossil matter shows all its interest and potential, in a perspective of Closed Cycles.

So CIRPS has carried out studies focused on: electrolytic (hydro, wind, PV) and thermal/biologic processes (biomass).

[1] CIRPS was founded in April 1988 as a Common Research Centre of Italian Universities. "La Sapienza" of Rome is the leading partner of the Consortium . Other Current members are: University of Viterbo, University of Cassino, University of Perugia, University of Torino, University of Sassari, University of Macerata, University of Palermo, University of Lecce.

A. Iacomelli (ed.), Renewable Energies for Central Asia Countries: Economic, Environmental and Social Impacts, 95–101.

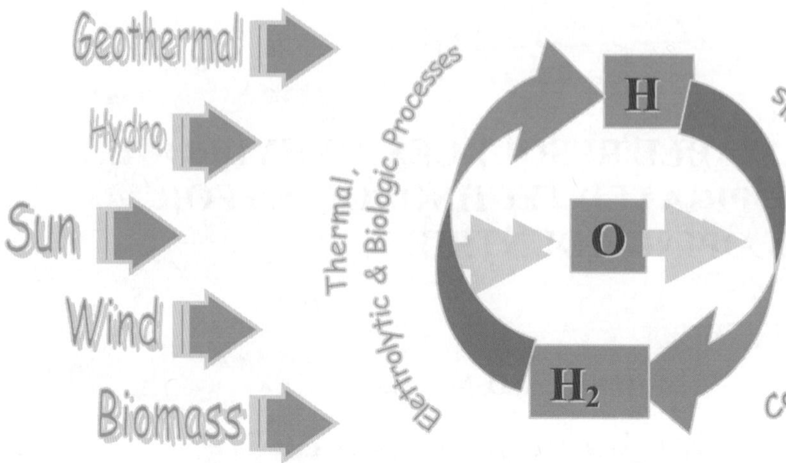

2. PHOTOVOLTAIC, MINI HYDRO AND WIND HYDROGEN PRODUCTION

One of the most promising methods of producing hydrogen in an environmentally clean way is the so-called solar hydrogen cycle. A model for a solar-hydrogen energy system for Angola has been developed by CIRPS.

The system is constituted by a Photovoltaic Module Array and a Fuel Cell Energy Storage Module.

There are the following characteristics: hydrogen production through electrolysis during the day on the base of the energetic surplus; low pressure hydrogen storage; use of hydrogen for electric power production during the nocturnal hours.

The system is based on the following components: Photovoltaic Module, Hydrogen Production Unit (electrolyser), Hydrogen Storage Unit, Conversion Unit Hydrogen-Electric Power.

The solar hydrogen cycle is currently a real alternative to fossil fuels, and its market penetration will require systems to be set up, such as the one described here.

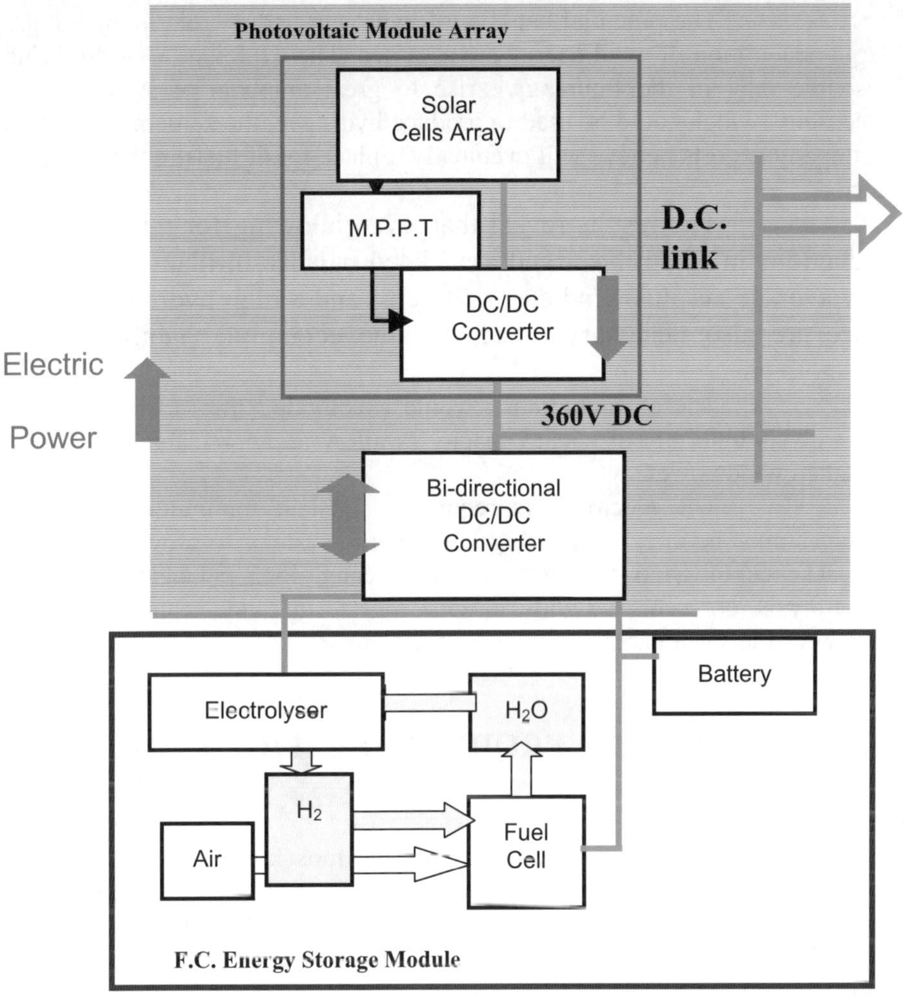

Wind-powered water electrolysis is envisaged as an important source of zeroemissions hydrogen in the future. Although it may appear very costly at a first glance, the use of surplus wind power in scenarios, where wind represents a large share of the electricity market or in off-grid systems, points to a different outlook. Hydrogen systems can help to overcome the problems arising in the electric system at high wind energy penetration or where there isn't a grid enable to cover wind uncontrollability. CIRPS studies the viability of hydrogen production via electrolysis using wind power in developing countries with high value of wind (like Peru: 5 to 6 m/s annual average).

The potential benefits of hydrogen and its role in enabling a large penetration of wind energy are assessed. The exploitation of this wind resource may in the future give rise to great amounts of surplus wind electricity, which could be used to produce hydrogen, the zeroemissions fuel that many experts believe will eventually replace fossil fuels in the transport sector.

The results reveal that, even allowing for relevant cost-reductions in electrolyser and associated balance-of-plant equipment, low average surplus wind electricity cost and a high hydrogen market price are also necessary to achieve the economic viability of the technology.

These conditions would facilitate the installation of electrolysis units of sufficient capacity to allow an appreciable increase in installed wind power.

The electrolysers require a constant minimum load, wind turbines must be integrated with grid or energy systems to provide power in the absence of wind. Perfecting sophisticated load balancing for wind electrolysis represents a significant challenge to the wide-scale deployment of renewable/sustainable energy infrastructure.

3. HYDROGEN PRODUCTION FROM BIOMASS: SUGARCANE[2]

The sugar cane, is one of the most interesting biomass for energy purposes because has high rate of energy produced per hectare (50-100 t/ha, CV 8-20 MJ/kg); rich typologies of waste at high energy content (leaves and tops, bagasse[3], molasses, etc); no production and transport costs, since the most of waste are produced and used in the factory: the only cost is due to the collection of leaves and tops from the field; and favourable characteristics of cultivation (warm-humid climates, small demand for pesticides and chemical fertilisers but a lot of water: 2000 m/year).

---

[2] COOPERATION ACTIVITIES CIRPS – PERU. General Framework Agreement Cooperation between the University of Rome "La Sapienza" CIRPS – CIDES (University of Chiclayo "Santo Toribio de Mogrovejo") and with the University of Piura. Such cooperation will be carried out on basis of equality and mutual advantage, reserving the possibility to define further fields of activity

[3] BAGASSE is the fibre of the cane obtained by milling and pressing the cane in sugar mills. It is used marginally in the energy sector.

CIRPS evaluated both the use of bagasse and barbojo[4] in order to produce hydrogen, and the environmental, social and economical effects. The aim is to develop a technical guidelines for the design of a small power plant (gasifier-fuel cells) with the highest efficiency, low cost and low pollutant emissions. The case study is a sugarcane mills in the north zone of Peru.

The power plant was assessed in the 2000; produces 905.300 t (80 t/ha.y) of sugar cane, 633.724 t (187 t/ha.y) of worked cane, and 64.640 t of sugar. The plant, situated near Trujillo on the coast in North of the Peru, burns the barbojo in the fields and uses old low efficiency boilers; thus it is not energetically self-sufficient, mainly for electric power.

The use of barbojo could became the power source for the sugar factories in advanced integrated biomass technologies for new energy carriers, like hydrogen. This can become one of the most promising and cheap options when advanced gasification-fuel cell technology is used.

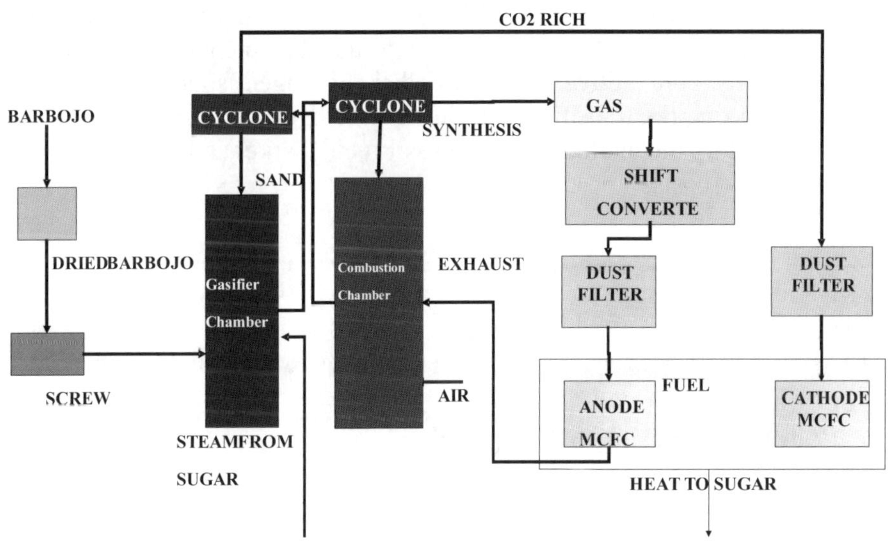

4. CONCLUSIONS

The sustainable development strength is in the weakness of current development:

4 BARBOJO is the sugarcane trash, the tops and leaves that are abandoned in the fields after the harvesting

- Impossibility to answer of the even more explosive environmental, economic and social problems.
- Finite (and now nearly achieved) environmental tolerance capacity.
- Finite fossil resources.

The choice of instruments to solve this problems all around the world represent the greatest opportunity for humane race.

This paper has shown that hydrogen production without polluting emissions is feasible using a renewable source of energy and with technical means that are currently available in the market.

Hydrogen has attractive features for renewables integration, primarily on account of its multi-functionality. Hydrogen acts as a storable 'energy carrier' that can be either converted back into electricity thus providing a balancing service to electricity generators and suppliers- or used as a 'zero emissions' fuel for other applications, such as transport.

The use of renewables-generated hydrogen in the transport sector will have a substantial contribution to the abatement of CO_2 and other emissions. Its competitiveness with other hydrogen sources like steam reforming during the transition to a carbon-free hydrogen economy will also be influenced by the evolution of the fossil fuels market and the reflection of environmental costs.

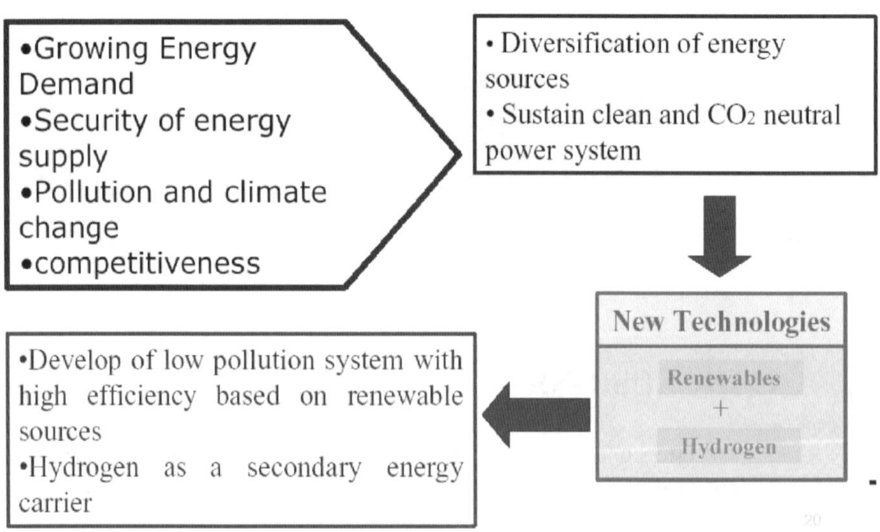

REFERENCES

[1] "Barbojo: a tool for greenhouse gases reduction". Proceeding of Sixth International Conference on Technologies and Combustion for a Clean Environmental, Clean Air. New Customs House, Oporto, Portugal. July 2002. Naso, Orecchini, Arroyo, Santiangeli.

[2] "Utilizzo degli scarti della canna da zucchero per la produzione di idrogeno in Peru". Proceeding of X Convegno Tecnologie e Sistemi Energetici Complessi. Genoa 2001. Naso, Orecchini, Arroyo, Santiangeli, Zuccari.

[3] "Produzione ed utilizzo energetico dell'idrogeno dagli scarti della canna da zucchero in Peru". Degree thesis – University of Rome "La Sapienza". 2002. Bocci.

[4] "Hydrogen production in Peru based upon sugarcane waste". Ph.D. Thesis. University of Rome "La Sapienza". 2002.Arroyo.

[5] The role of hydrogen in high wind energy penetration electricity systems: The Irish case. A. Gonza´lez, E. McKeogh, B.O. Gallachoir. Department of Civil & Environmental Engineering, UCC, College Road, Cork, Ireland, 29 July 2003.

Chapter 9

THE CONCEPTION OF THE USE OF RENEWABLE ENERGY SOURCES AND THEIR ROLE IN THE ENERGY BALANCE OF UZBEKISTAN

T.P. Salikhov, T.H. Nasyrov

Institute of Power Engineering and Automation of Uzbekistan Academy of Sciences
Akademgorodok, 29 Khodjaev str.
700125 Tashkent, Republic of Uzbekistan

Prominent international organizations in the energy field, such as the International Energy Agency, the American Society of Electric Engineers, predict wide involvement of renewable sources of energy in the global energy balance already in the nearest future and state that their share in energy balance can reach 40 % in 50 years. In this connection it seems to be reasonable to analyze the present readiness of Uzbekistan for introduction of renewable energy sources.

The Kyoto Protocol was ratified by Uzbekistan in 1999 approving the readiness to decrease the emission of greenhouse gases and develop renewable energy. According to Kyoto Protocol Uzbekistan has no quantitative restrictions on the greenhouse gas emissions and can only participate in the activity within the framework of Clean Development Mechanism. Uzbekistan maintains continuous monitoring over greenhouse gas emissions in 21 economic sectors. The data on gas emissions with direct greenhouse effects are presented in the Table 1.

A. Iacomelli (ed.), Renewable Energies for Central Asia Countries: Economic, Environmental and Social Impacts, 103–121.

Table 1. The dynamics of greenhouse gas emissions in Uzbekistan (in million tons of CO_2 equivalent) [1]

Gas	1990		1999		2000	2000
	Emission	Share %	Emission	Share %	Emission	To 1990г. %
Carbon dioxide CO_2	114,6	70,2	104,7	65,3	109,5	95,5
Methane CH_4	37,7	23,1	46,6	29,0	47,8	126,8
Nitrous oxide N_2O	10,9	6,7	9,1	5,7	13,3	122
Total	163,2	100,0	160,5	100,0	170,6	104,5

The trend of declining combined emissions of greenhouse gases in evidence in Uzbekistan over 1990-95 gave way in 1996 to increase in total emissions. Emissions of not only methane, but also carbon dioxide have begun to grow. In 2000, greenhouse gas emissions in Uzbekistan worked out to 7.0 tons per capita, which was above the world's average and equal to comparable emissions in France and Spain. The exhaustion of opportunities to diversify fuel consumption (with gas in Uzbekistan accounting for 80 % of all fuels used) and the fact that emissions in 2000 were up on those in 1990 lend extra difficulty to the task of decreasing such emissions, and imparts top priority to the program of measures to promote energy savings and to steadily expand the share of renewables in the nation's energy mix [1-4].

The total potential of renewable energy sources in Uzbekistan is almost 6.8 trillion toe and their technically feasible potential is estimated at 179 million toe, of which only 0.6 million toe, or 0.33 % has been put to use.

Table 2. Potential of renewable energy sources in Uzbekistan [2].

Parameter	Total	Energy, Mtoe		
		Hydro	Sun	Wind
Technical potential, M t.o.e.	179,0	1,8	176,8	0,4
Put to use	0,6	0,6	-	-
CO_2 emissions potential, Mtoe	447,5	4,5	442,0	1,0
Put to use	1,5	1,5	0	0

The above table makes it clear that if the technical potential of renewable energy sources in Uzbekistan is tapped in full, it will replace fossil fuels producing CO_2 equal to 447.5 million toe when burnt. It is apparent, that the potential for renewable sources of energy in Uzbekistan is large and the opportunities for project realization within the framework of Clean Development Mechanism are considerable. However they are only the initial preconditions favorable for the development of renewable energy in Uzbekistan. The full forecast of development of alternative energy is only possible in the context of the energy policy of the country and its conformity to the tasks of the development of the fuel and energy complex and the national economy. Only having estimated the condition of the fuel and energy complex and the problems of the energy policy, institutional and legislative base for renewable energy sources and economic opportunities it is possible to predict the development of alternative energy.

Uzbekistan is presently considered to be a country in transition. The strategy of transition to a market economy has been developed by its President, I. A. Karimov. It is based on five principles, namely: a depoliticized economy, a regulatory role for the government, the rule of law, efficient social policies, and consistent reforming on a phased basis. The developed way of transition to market relations is aimed at forming a socially - oriented market economy to adequate to the interests, conditions and national features of the country.

Structural policies in the fuel and energy complex during 1991-2003 were pursued amid intensifying economic reforms and aimed to achieve the priority objectives of the energy strategy designed:

- to create a reliable resource base and to support strategic economic sectors;
- to increase the efficiency of energy resources usage and to furnish essential conditions for the implementation of energy saving measures as required in order to preserve the nation's energy independence and its export potential;
- to sustain the energy sector's financial stability and to attract additional investments in its development, to revamp the legal and regulatory environment and upgrade the fiscal system with due regard for the special aspects of pricing in the energy sector and its relations with related sectors, to consistently cultivate the competitive environment in the energy sector by creating full-blooded market players and adequate market infrastructure.

According to the strategy for transition to a market economy, as proposed by the President I. Karimov, realization of the energy policy in

Uzbekistan was carried out gradually. The primary tasks at the 1st stage of the energy policy in Uzbekistan during the period of transition have included:

a) maintenance of energy independence;
b) achievement of full scale access of the population to natural gas;
c) keeping energy prices low so as to maintain living standards and the competitiveness of local producers at creditable levels.

Rich in primary energy resources, Uzbekistan has successfully solved the above problems of the first stage of the energy policy. Over the years of independence it has trebled its crude oil output from 2.8 million tons in 1991 to 7.5 million in 2001, natural gas production has gone 50 % up – from 41.9 billion cubic meters in 1991 to 63.1 billion in 2003. Uzbekistan turned into a net exporter of natural gas as early as in 1995, while its import of crude oil, which had amounted, in monetary terms, to USD 485 million that year, dwindled to zero in 1996-97. At this stage the gradual process of privatization of the energy sector started. Thus in 1994 the Resolution of the Cabinet of Ministers № 290 of 09.07.1994 « On the Experimental Privatization of Particular Petrol Stations Selling Petrol to the Population in the City of Tashkent for Cash » was issued.

Both the annual generation and demand for electricity have been on the increase since 1996. Table 3 below shows the development of electric power supply and demand between 1995- 2001.

Table 3. Electric Power Generation and Consumption in Uzbekistan

	1995	*1999*	*2000*	*2001*
Electric Power Consumption, GWh	*42020*	*43015*	*44017*	*45011*
Electric Power Generation, GWh	*47453*	*45300*	*46840*	*47927*

Source: Energy Balances of non-OECD countries 2000-2001, IEA Statistics, 2003 Edition

As can be seen from the above, Uzbekistan achieved self – sufficiency in fuel in 1995 and became fully self – sufficient in energy in 1996-1997.

As far as the task of making natural gas supplies more readily accessible to the population is concerned, it should be said that a really great deal has been done towards this end over the years of independence, with more than 3.5 million households, or *95% of the population,* now enjoying *access to the natural gas distribution network* and an extra 720000 households using liquefied gas these days as a result of those efforts. The

Uzbek Government intends to get the remaining 5% of the population connected to the gas distribution network as well within the next three years.

Efforts to keep energy prices low were required at the 1st stage in the nation's energy policies, because Uzbekistan has decided against any shock therapy during its transition to a market-based economy. Its Government has opted instead for the evolutionary path of economic reforming, the result being that the reform drive in the energy sector as a basic national industry has proceeded gradually, on a phased basis. It is planned that energy prices will rise as it makes further headway not only in the energy sector, but in the economy as a whole.

The 2nd stage in Uzbekistan's energy policies has seen a plan-based reforming of the various branches of the energy sector with a gradual implementation of market-based mechanisms there. The reform in the oil and gas industries started in 1998, in the power industry in 2001, and in coal mining in 2002. The economic reform in the sector has been accompanied by institutional change and the provision of the required legal and regulatory framework. The institutional change started with the establishment of Uzbekneftegaz National Holding Company in 1998, which was made responsible under Presidential Decree No. 2154 dated November 11, 1998 for managing the country's entire oil and gas sector. The organization and operations of Uzbekneftegaz were defined in Cabinet of Ministers Resolution No. 523 dated December 15, 1998. The company has eight subsidiaries.

The commencement of that kind of institutional change was accompanied in the same year of 1998 by the launch of building work on the Shurtan plant to produce polyethylene and liquefied gas, Central Asia's largest facility of this kind. The factory, which went on stream recently, is capable of annually producing 125,000 tons of polyethylene, 137,000 tons of liquefied gas, and 37,000 tons of light gas condensate. Likewise in 1998, reconstruction work went ahead in cooperation with the Japanese Mitsui on the Fergana oil refinery, which manufactures gasoline, diesel fuel, and jet fuel to world standards.

In order to ensure full and reliable supplies of natural gas to industrial centers, an underground storage for 1,800 million cubic meters of gas was built at Khodzhaabad in the Fergana Valley in Uzbekistan in 1999.

Therefore, the years between 1991-2000 saw the Uzbek oil and gas industries fulfill the following three paramount tasks:

a) The nation's self-sufficiency in fuel was achieved;
b) A marked increase in the output of export-oriented products of gas processing and oil refining operations and in natural gas exports as a result of the sector's restructuring; and
c) A legislative basis put in place as required to attract foreign investments in the oil and gas sector.

The Uzbek power grid has an installed capacity of 11.2 million kilowatts, and comprises 9 thermal power stations with 63 turbo-generators having a total generating capacity of 9.8 million kilowatts (or 87.5% of the total), and 28 hydroelectric power stations with 67 hydraulic turbine generators having a total generating capacity of 1.4 million kilowatts (12.5%), as well as three departmental electric power stations with a total generating capacity of 319,000 kilowatts. The power distribution network totals 231,000 kilometers of lines, and has a transforming capacity of 44 million kilovolt-amperes.

Since developing a socially oriented economy is a priority objective of the ongoing reforming, the projected reform of the existing smooth running, but centralized power sector requires a balanced, consistent, and prudent approach. Before 2001, the sector was called upon to provide steady, uninterrupted supplies of electricity to all other branches of the national economy by making use of its advantages as a vertically integrated monopoly. However, with the reforming of all spheres of social and economic life, the unduly centralized system for managing the generation, transportation, and distribution of electric energy no longer makes it possible to cope with the tasks of making such operations more efficient and cost-effective. This is why economic reform went under way also in the power sector in real earnest in 2001 in line with Presidential Decree No. UP-2812, "On Furthering Reform in the Electric Energy Industry in the Republic of Uzbekistan," dated February 22, 2001.

The Decree has set the following goals:

- advancement of market-geared reform in the sector;
- its demonopolization;
- higher economic efficiency of enterprises operating in the industry;
- wider-scale privatization with the involvement of domestic and foreign investors;
- improvement of quality and security of electricity supplies to consumers.

The reform drive in the sector is to prioritize the following:

- consistent demonopolization of energy enterprises;
- looser government regulation, and better conditions for competition among different electricity distributors;
- progressive fostering of wholesale and retail markets of electric and thermal energy;
- equal opportunities for all competitors, including equal

access to power transmission lines;

- implementation of efficient arrangements and methods for developing coal fields, and broader reliance on coal as fuel in power generation;
- extensive attraction of domestic and foreign capital in the modernization and re-equipment of energy facilities;
- stage-by-stage transformation of thermal power stations, combined heat and power plants, and regional power networks into joint stock companies with the continued controlling equity participation of the state-owned Uzbekenergo;
- transformation of construction, installation, and repair enterprises into joint stock companies which will continue to be at least 25% controlled by the state.

The Decree has also effected some institutional change by way of demonopolizing and improving the management system in the sector. The Uzbek Ministry of Energy and Electrification has been abolished, with the state-owned Uzbekenergo public joint stock holding concern organized on its basis to incorporate also the Ugol joint stock coal-mining company. Other entities resulting from the measure have included a high-voltage network subsidiary (Uzelektroset) to transmit, and regional distributors to allocate, Uzbekenergo's electricity.

The Uzgosenergonadzor state agency for supervision over the electric power sector under the Uzbek Cabinet of Ministers, initially established as the government regulator for the electric and thermal energy industries and coal mining, has been converted into the Uzgosenergonadzor state inspectorate under the Uzbek Cabinet of Ministers in accordance with the latter's Resolution No. 96 dated March 1, 2004. An immediate follow-up on the above presidential decree, the Cabinet of Ministers' Resolution No. 93 of February 24, 2001, "On Measures to Organize the Operations of Uzbekenergo State Joint Stock Company," has detailed practical measures to implement that Resolution on power sector reform and presented further steps to refurbish and privatize Uzbekenergo. They have included the following:

1) Phased corporization of thermal power stations and regional power distribution networks with 40% interests to be offered to investors;
2) Sale of more than 75% of the shares to investors during the privatization of design, construction, installation, and repair enterprises;
3) Transfer of state-owned equities and shareholdings in thermal power companies to appropriate utility and maintenance

associations for trust management; and

4) Privatization of Uzbekenergo social infrastructure facilities.

Step-by-step reforming is the leading principle of the nation's energy policies. This is why the fulfillment of the overriding task - that of becoming independent in the energy field - should be followed by the attainment of national energy security. As already noted above, the share of gas in the country's fuel budget is exceedingly large and this fact detracts from its energy security. Therefore, fuel diversification is among the key aspects of the energy security issue. It would be sensible for Uzbekistan, which boasts ample reserves of brown coal, to expand the share of that resource in the national fuel budget. For this reason, coal sector reform has come as the natural next phase in the country's energy policies. Total coal production at initial stages in the reform process tended to decline, but the subsequent implementation of measures towards greater energy diversification has made for further and steady gains in coal output. The share of the power sector, namely: electric power stations, in the combined consumption of coal as fuel has reached 80%, with other users accounting for the remaining 20%.

The table below offers information about the production and use of coal in Uzbekistan.

Table 4. Coal production and consumption in Uzbekistan over 1992-2000

	1992	*1995*	*1996*	*1997*	*1998*	*1999*	*2000*
Production, Mtoe	*1.66*	*1.08*	*1.00*	*1.04*	*1.03*	*1.05*	*0.91*
Consumption, Mtoe	*2.18*	*1.07*	*1.20*	*0.98*	*1.03*	*1.02*	*0.88*

Sources: Energy Balances ofnon-OECD countries 2000-2001, IEA Statistics, 2003 Edition

The reform drive in coal mining commenced in 2002 with the issue of Cabinet of Ministers Resolution No. 196 dated June 2002, which has set out the Development Program for the Coal Industry in Uzbekistan for the Years 2002-10. The Programme provides for stage increase in coal output from 2.7 million tons in 2001 to 9.4 million in 2010 and in the share of coal used for power generation in the national energy balance from 4.7% in 2001 to 15% in 2010. The aggregate output of primary fuel and power resources remain very substantial, being increased by 14.3% from 1991 to 2003.

Uzbekistan is among the principal producers of oil and gas in the Central Asian region. Its fuel budget in 2000 consisted of 83.4% of natural gas, 13.7% of oil and condensate, 2.2% of coal, 0.7% of hydropower, and 0.03% of mine gas in coal mining.

Gas production in 2003 came to 58.1 billion cubic meters. It is concentrated on 12 fields and is carried out primarily by the oil and gas companies Mubarekneftegaz, Shurtanneftegaz, and Ustyurtgaz. Most of the gas processing is done by two facilities - the Mubarek gas processing mill and the newly commissioned Shurtan gas and chemicals complex. A total of 43% of all gas used is consumed by households, 30% is burnt to generate electricity, and the remaining 27% is put to industrial uses. The growing share of the population in the combined consumption of natural gas (with the number of households with access to gas supplies over the years of reform having almost doubled) demonstartes the social concern of Uzbekistan's energy policies.

Oil and condensate output in 2003 amounted to 7.2 million tons. Production operations are maintained by Uzgeoburneftegazdobycha, Mubarakneftegaz, Shurtanneftegaz, Dzharkurganneftegaz, and Mingbulakneftegaz. Following the construction of a new oil refinery in Bukhara (capable of processing 2.5 million tons of crude oil per annum) and the reconstruction of similar facilities hi Fergana and Altyaryk, Uzbekistan has enough manufacturing capacity to refine 11 million tons of oil yearly. The production of motor gasoline grew by 18% over 1995-2003. 65% of the total amount of its consumption on the domestic market is realised by the population, 12% by agriculture, and 23% by transport and other economic sectors. More than 82% of the gross consumption of heavy fuel oil falls on the power sector (i.e. electric power stations), with the other industries being responsible for the remaining 18%. The largest user of diesel fuel is agriculture (63.5%).

Coal mining in Uzbekistan, as already reported above, is done by Ugol, a joint stock holding of five producing companies. Combined coal production at the initial stages in the reform process tended to decline, but the measures adopted to diversify the fuel budget as soon as possible have reversed the trend, making for a steady growth of the output of this natural asset. The power industry, namely: electric power stations, accounts for 80% of the entire national use of coal fuel, with the other users consuming the remainder. A factory having the rated production capacity of 200,000 tonnes of coal briquettes per annum has been built at the Shargunskaya mine. The binding agent employed in the manufacturing process there is petroleum bitumen. Top-quality coal from Baisun can be used as a valuable chemical starting material (in the production of coked briquettes, adsorbents, liquid fuel, etc.).

The Uzbek power system has the potential to generate 56-58 billion kilowatt hours of electricity. Gross electricity generation in 2002 reached 49.3 billion kilowatt hours.

The power system in Uzbekistan is rallied round the Syrdarya, Tashkent and Novosibirsk-Angren thermal power stations and the Navoi district power plant. Together, they carry 37 generators with a unit power rating of 150 to 300 megawatts each. Thermal power stations are responsible for around 88% of the gross electricity output, while the rest comes from hydroelectric power stations. Natural gas is the main component of the boiler and furnace fuel balance. The Uzbekistan power system makes an element of the Unified Energy Systems of Central Asia, which also incorporates the power grids of Turkmenistan, Tajikistan, Kyrgyzstan, and Southern Kazakhstan. Nearly one-half of the Unified Energy Systems' generating capacity is found in Uzbekistan. Over 52% of all electricity is consumed by industry, 25.6% by agriculture, and 12%-15% by households.

Uzbekistan also generates some 200,000 TJ of thermal energy annually, and uses about 5 Mtoe, or 10% of the aggregate national fuel consumption for this purpose. A total of 34% of the thermal energy output goes to heat homes and supply them with hot water. Almost 17% is used for the social sphere, with another 30% used by industry and 19% by the energy sector.

Therefore, Uzbekistan has been able over a relatively brief time to ensure the stable advancement of its vast fuel and power sector. However, in order to sustain the country's self-sufficiency in energy and its export potential, it is extremely important to improve the efficiency of energy uses and set the necessary conditions for implementing the energy conservation policy formulated by the Uzbek Government in 2002.

A comparison of energy/output ratios in Uzbekistan and developed Western countries, which have been vigorously pursuing energy-saving policies and providing clear-cut legislative regulation for the production and use of energy resources, reveals that the specific energy intensity of the Uzbek GDP is 2-2.5 times higher and stands at 1.05 kilograms of oil equivalent per US dollar (based on 1995 PPP). The principal reasons why the energy/output ratio in the Uzbek GDP remains high include the existing energy-intensive industrial production structure, and the technological inadequacy of fixed production assets resulting in the amount of energy required per unit of output in some processes in mechanical engineering, metal-making, the chemical industries, and certain other sectors being substantially larger than comparable figures in developed economies. This makes energy-saving the top priority in the state's energy policies all the way until 2010. According to expert estimates, it is technologically and commercially feasible even under the existing pattern of energy consumption to cut the current use of energy by 15-18 million toe, or 23%-28% through saving measures.

The largest savings can come from improvements in the residential sector (51.2%), the power sector (8.5%), the oil and gas industries (10.3%) and other economic sectors (30%).

The ongoing ***Program for Energy Conservation in the Republic of Uzbekistan in the Years Until 2010*** has been prepared in accordance with Article 12 of the Uzbek ***Law "On Rational Energy Uses"*** and Cabinet of Ministers Resolution No. 517 dated December 29, 2000, with due regard for the latter's decisions of February 14, 2002 and with the participation of national ministries and agencies, associations, unions, and organizations, as. well as the Council of Ministers of the constituent Republic of Karakalpakistan, regional hokimiyat administrations, and the Tashkent city council. The relevant industrial and regional energy-saving programs have been elaborated with the involvement of 48 national ministries, agencies, companies, associations, and organizations and the authorities of all of the 14 regions, including the regional hokimiyat administrations, as well as the Karakalpakistan Republic Council of Ministers and the Tashkent city council. Lists have been drawn up of basic energy saving measures to be taken by the fuel and power sector and by the end users, with the relevant computations made on the basis of the corresponding repayment periods and specific costs required. Corresponding saving priorities for the fuel and power industries have been set.

Under the National Energy Conservation Program, average annual savings of fuel and power resources in the whole of Uzbekistan are expected to total 11.08 million tons of oil equivalent (Mtoe).

The largest savings are anticipated to be made in the energy sector, including Uzbekneftegaz (1.148 Mtoe) and Uzbekenergo (0.324 Mtoe). Industry is to save a total of 1.882 Mtoe (Uzkhimprom - 1.749 Mtoe). Average annual savings in the agribusiness sector are to total 0.403 Mtoe. Another 93000 tons of fuel and power resources in terms of their oil equivalent are to be saved in the production of consumer goods .and the trading industry. The comparable figures in the construction industry, communications, and the utilities sector are 0.125 Mtoe. In the residential sector, they amount to 6.067 Mtoe. Absolute priority among energy-saving measures both in the fuel and power sector and among energy users goes to the so-called no-cost measures, i.e. improvements to arrangements already existing, introduction of incentives for energy efficiency, and efforts to achieve elementary order in energy uses. Such measures can tap 5% of the entire energy conservation potential at enterprises. Users' supply with gas and electricity meters are important steps that pay back for themselves fast. Non-centralized heating systems in the residential sector help to save a great deal of thermal energy where inefficient boilers are currently used. While the installation of meters cannot per se save any energy, this equipment makes it possible to reduce commercial losses during the transportation and use of fuel and power by identifying and locating the heaviest loss points. Where meters are supplemented by energy use regulation systems and other energy-conservative facilities and arrangements, the combined effects can account for up to 50% of the entire energy-saving potential.

Emphasis in energy saving efforts is on low-cost activities promising quick returns (with a payback of 2 years or less), measures fully paying back for themselves within 3-4 years, and those with payback of 5-7 years. Low-cost activities include, for example:

- measures designed to cut back on idle runs by production equipment;
- identification of optimal operating modes for technological facilities, and other production mode rationalization;
- indoor temperature regulation depending on outside air temperatures;
- development of regulations on financial incentives to employees for higher efficiency and achievements in implementing energy saving programs;
- energy audits with a view to devising rational energy use modes;
- elaboration of scientifically validated use tariffs for fuel and power resources by the type of production manufactured;
- plotting of rational road haul and railroad itineraries;
- electricity savings by making the most of natural lighting;
- better insulation to decrease heat losses;
- market saturation with economy cars to save oil products in the consumer sector;
- prevention of fuel and lubricant leaks from valves, painting of fuel tanks, and mounting of breathers;
- use of distribution apparatus, and organizational measures to improve energy efficiency (including personnel training, and organizational streamlining).

Such low-cost activities are expected to yield 0.107 Mtoe in fuel and power savings. Steps capable of paying back for themselves within two years include, but are not limited to:

- installation of condensate discharge pipelines, and implementation of computer-controlled energy use monitoring and control systems;
- use of Alfa-type electric meters;

- burning equipment modernization, and replacement of existing boilers by more advanced facilities;
- preparation of optimum mode cards for boilers, and waste heat recovery through the installations of heat-exchanging units at medium- and low-pressure pipelines;
- installation of gas, heat, and other energy meters;
- automation of power units at pumping and compressor stations;
- change-over to local heating systems;
- use of variable-speed drives in engines;
- cuts in water losses during the watering of irrigable lands (through such methods as gravity flooding, film-based technologies; drip irrigation, etc.), and replacement of diesel pumps by electrically-power pumps for intra-farm irrigation purposes;
- reduction of losses in farm produce (through the outfitting of storage facilities, construction of refrigerators, and use of specially-equipped vehicles);
- optimized operating modes for mills at flour milling enterprises, and installation of extra condenser-type compensating units;
- increasing use of modern, lower-power lamps for lighting purposes;
- maximum reliance on secondary thermal energy by rationally using waste-gas heaters at superheaters, and switch-over from “Vet" to "dry" production modes in the manufacture of cement.

Total savings from the above measures should add up to 1.528 Mtoe. Further steps along these lines which can bring full-scale returns within three to four years include:

- construction of modularized gas-turbine thermal power plants of the GTES-4 type, and reconstruction of GT- 750-6 Avrora gas-pumping units with the replacement of centrifugal pumps by double-stage pumps to increase the equipment's unit capacity to 8 megawatts and its performance factor to 34% (Uzgeoneftegazdobycha and Uztransgaz);
- modernization and repair of thermal power plant facilities Uzbekenergo), and replacement of Bratek and NR-18 type boilers by more economical KV-0,25 and AO GV 46,5 facilities (Uzbek State Committee on Geology);
- replacement of steam-operated facilities by water-driven ones (Uzeltekhprom);
- reconstruction and modernization of heat use systems

(Uzkhimprom);
- better thermal insulation for buildings;
- modernization of a steel-making furnace, and installation of automated drying stands (Uzmetkombinat);
- phasing in of electric drive frequency regulators at enterprises;
- reconstruction of steam conduits and heat supply networks;
- vehicle conversion to run on gas, and construction of solar power plants;
- provision of consumers with cold and hot running water, heat, and gas meters;
- pump replacement at water trunk pipelines.

The above measures with recoupment periods of three to four years can save fuel and power amounting to 5.077 Mtoe. Energy-saving measures capable of paying off within five to seven years include:

- modernization of GTN-6 and GT-6-750 gas turbines and their replacement with gas pumping facilities having a performance factor of 35%-37% (Uztransgaz);
- Angren mine modernization (Ugol);
- updating of production equipment, implementation of inter-summer air conditioners and regulated electric drives, and the own manufacture of electrically-powered compressor stations to replace diesel compressors (Uzmashprom);
- installation of mini-boilers to heat four residential townships (Uzeltekhprom);
- reconstruction and modernization of technological equipment (Fergana-based Azot, and Navoiazot);
- construction of steam and gas turbine plant (Azot, Navoiazot, EKhP, Ammofos, and Samarkand chemicals factory);
- implementation of non-centralized heat supply systems;
- compressor station modernization;
- transport converter implementation;
- construction of solar heaters;
- cotton mill modernization and new construction (Uzkhlopkoprom), and reconstruction and retooling of existing enterprises (Maslozhirtabakprom);
- pipeline replacement and heat supply pipeline insulation;
- implementation of a computer-controlled system for outdoors lighting;
- replacement of deep-well pumps at water intake facilities;

- vehicle conversion to run on compressed gas;
- implementation of a demonstration project providing for the implementation of an energy-efficient zone for automated control over heat uses.

All in all, energy saving measures with payback periods of five to seven years can yield fuel and power savings amounting to 2.765 Mtoe. There also exist plans for measures with payback period of longer than seven years, which can also be seen as indirect energy savers. They include:

- implementation of cryogen-based technology for processing gas, recycling associated petroleum gas and rock decay gas at the Kokdumalak field (Uzgeoneftegazdobycha);
- updating of schemes for and construction of power and steam supply systems (Uzneftepererabotka), and installation of a 5-10 kilowatt NW wind power plant (Uztransgaz);
- equipment modernization and repair (at the Syrdarya, Novosibirsk-Angren, Takhiatash, and Navoi thermal power stations and the Fergana, Mubarek, and Tashkent district electric power plants);
- generator building and installation (Uzmashprom);
- production of frequency converters, and triple-phase active energy meters, and an accumulated register for tariff regulation (Saben);
- commencement of production operations to manufacture low-power lighting lamps (Foton), and construction of small hydroelectric power stations at water works (Ministry of Agriculture and Water Management);
- increasing use of diesel and gas engines on motor vehicles;
- implementation of solar plants for hot water supplies;
- and harnessing of alternative and renewable energy sources.

As one can see, at a present stage of development of the fuel and energy complex the state aims at introducing energy saving measures and renewable energy sources in Uzbekistan. In other words, having achieved sustainable functioning of traditional power, the state begins supporting the development of non-traditional and renewable energy, keeping to the main principle of gradual reforms. Until now the development of renewable energy in the republic was carried out within the framework of individual projects and limited by academic researches. There are a few organizations in Uzbekistan specialized in the field of renewable energy sources: the joint-stock company " Hydroproject ", Hydrometeorological Institute, the Agency for the Transfer of Technologies, Institute of Power Engineering and Automation, Physical Technical Institute of the Uzbekistan Academy of Sciences, the Center of Scientific and Technical Marketing Research, Tashkent State Technical University, the Center of Science and Technologies. Within the framework of the project "INCO-Copernicus" of the European Community with the participation of the Institute of Power Engineering and Automation, the Center of Scientific and Technical Marketing Research and specialists from France and Portugal, a pilot system of power supply with wind energy device with a capacity of 3 kW and a solar photo-electric unit with a capacity of 6 kW was created to supply energy for a relay station in Charvak settlement. Research proved the expediency and the efficiency of combined use of wind and solar energy in Charvak area for power supply. Besides, the Institute of Power Engineering and Automation carry out estimations of the energy potential of wind, hydropower resources, solar energy in Uzbekistan. Also the recommendations for use of various types of installations on the basis of renewable energy sources are developed by the Institute for power supply of consumers in various geographical zones of Uzbekistan. The firm " Kurilishgelioservis" at the Physical Technical Institute produces solar collectors for all-the-year-round hot water supply. Produced systems of hot water supply are twice cheaper than foreign devices of the same productivity. According to the estimations of the scientists of Power Engineering and Automation and Physical Technical Institutes for achievement of 10 % of economy of fuel and energy resources in hot water supply systems it is necessary to reach operation of solar collectors with the total area of 2 million in m^2 and annual production of 200 thousand pieces. It is to be noted that there is the experimental plant producing solar elements from mono-crystal silicon up to 10 kW per year in Physical Technical Institute. Also a solar energy system on the basis of Stirling engine with a parabolic concentrator has been developed. The Agency for the Transfer of Technologies is establishing solar photo-electric systems in a number of farms in the Samarkand area, and is planning to create the combined installation using wind and the sun energy for water desalination in Karakalpakstan.

From the point of view of resource and ecology parameters the solar energy should be involved in fuel and energy balance of the republic. Unfortunately, from the economic point of view large-scale use of solar energy will demand significant financial expenses. At the same time, there is another argument for the benefit of solar energy development in Uzbekistan. The share of gas is extremely high in the structure of the country's fuel balance and forms 83 %. But relying on the only one energy carrier weakens its energy safety. Therefore a question of diversification of fuel is one of the major problems. In particular, it is expedient for Uzbekistan to have significant stocks of brown coal and non-polluting solar energy to increase their share in the structure of fuel and energy balance. It is to be noted that Uzbekistan possesses a large total potential of solar energy which by various estimations constitutes from 71362,2 million tons of equivalent fuel (tef) up to 76459.5 million tef and its technical potential ranges 247.52– 265,1 million tef a year. Taking into account, that the annual need of the republic for energy resources makes about 65 million t.o.e , it is apparent, that the technical potential of solar energy exceeds the demand several times.

Use of solar energy for reception of thermal and electric energy in Uzbekistan may be accomplished on several directions:

1) Reception of low-potential heat with its subsequent use in systems of hot water supply, i.e. transformation of solar energy into heat.
2) Direct transformation of solar energy into electricity with photo-electric systems.
3) Thermodynamic transformation of solar energy into electricity and high potential thermal energy. Here the issue is to create solar-fuel power stations with a large number of distributed parabalo-cylindrical receivers - converters of the concentrated solar radiation into thermal energy of heat-carrier, which is then used for producing steam with a view to receiving subsequently the electric power or thermal energy. In such schemes 75 % of the generated electric power is provided with solar energy and only 25 % of electricity is due to the use of traditional fuel. Application of similar technology in Uzbekistan could provide significant economy of natural gas, lower environmental contamination and increase the export potential of gas. Now the cost of service and operation of traditional power stations is 1.5 cents for 1 kWh, and the cost of 1 kWh of the electric energy generated by photo-electric devices is over 20 cents. But the cost of 1 kWh of electricity produced on the basis of solar energy can be reduced down to 8.7 cents. Therefore, it is necessary to combine efforts of economists, scientists, engineers for counting the economy of solar energy with all factors in view and then make decisions on its development in Uzbekistan. At the same time it is to be noted that the organization of solar collectors production in the systems of hot water supply is already economically expedient now.

There are good prospects for hydropower engineering in Uzbekistan. The established capacity of hydroelectric power stations existing in Uzbekistan makes 1700 MW. The average annual production of the electricity on them is above 6 billion kW a year though hydro resources of the republic allow increasing several times the production of the electricity with hydraulic energy. The matter is that the technical potential of more than 650 rivers flowing on the territory of Uzbekistan allows one to build 1100 hydroelectric power stations with total capacity of 207 kWh and with total production of electricity of 1532 million kWh a year. In addition using the natural water-flows according to Hydroproject joint-stock company estimations", it is possible to build 140 hydroelectric power stations on artificial water-flows of Uzbekistan with the established total capacity of 1726 MW and with the total production of electricity of 7213 kWh a year. In estimating the technical potential of artificial water-flows the creation of hydroelectric power stations on them with capacity from 10 up to 30 MW was proposed, however the creation of micro-hydroelectric power stations up to 100 kW and small hydroelectric power stations from 100 up to 10 MW practically was not taken into account even though such hydroelectric power stations are widely used in the Western countries. The construction of micro hydroelectric power stations is relatively inexpensive and there is no demand for their expensive and skilled service. They also can be widely used in Uzbekistan not only for electricity supply of rural areas, but also for their heat supply, hot water supply and heating. In 1995 the Government Decree « On the development of small and medium-sized hydroelectric power stations in the Republic of Uzbekistan » was passed. According to the program it is planned to construct 141 hydroelectric power stations including 15 till 2010 which after installation will generate 1300 million kWh of the electric energy annually.

Thus an issue of development of alternative energy in Uzbekistan is not technical, but an economic one. Despite the fact that the cost of energy devices on basis of renewable energy sources at a level of traditional power systems are expensive, the development of generating capacities on base of renewable energy sources demands significant financial expenses. Now, when the republic faces problems of further deepening of the reforms in the fuel and energy complex, the large-scale development of alternative energy in Uzbekistan with its further involvement into fuel and energy balance of the country is not economically reasonable yet. At the same time it is necessary to use actively small energy devices on the basis of renewable energy sources for power supply of the countryside and remote areas suffering the deficiency in electricity and heat. In order to support such projects limiting greenhouse gas emissions it is expedient to find financial means actively using the above mentioned Mechanism of Clean

Development. On the other hand, it is necessary to prepare a society for the perception of innovations, actively promoting energy saving technologies including renewable energy sources. It will allow forming energy saving culture in consciousness of people, realizing results of anthropogenic influence and preparing the conditions for transition to energy saving way of economic development.

REFERENCES

1. The Bulletin №4 of Central Asian Hydrometeorological Research Institute named after V.A.Bugaev « *Priority technological needs of Uzbekistan in the field of reduction of greenhouse gas emissions and mitigations of negative consequences of climatic changes* », Tashkent, 2002.
2. « *The First National Report of Republic of Uzbekistan on the United Nations Frame Convention on Climate Change* », Tashkent, 1999.
3. « *The First National Report of Republic of Uzbekistan on the United Nations Frame Convention on the Climate Change* », the Phase 2, Tashkent, 2001.
4. The Bulletin №2 of Central Asian Hydrometeorological Research Institute named after V.A.Bugaev «*The basic results of greenhouse gas emissions cadastre*», Tashkent, 2002
5. Kenisarin M.M. « Power engineering of Uzbekistan: conditions and problems », Journal of « The Central Asia and Caucasus » №2 (32), 2004.
6. Salikhov T.P. «Energy Charter Protocol on Energy Efficiency and Related Environmental Aspects. The Regular Review - Uzbekistan 2004», Secretariat of the Energy Charter, Brussels, 2004.
7. Salikhov T.P., Abaturov V. « Stages and results of energy strategy realization in Uzbekistan », Journal of « Economic Review », October, 2004.
8. Salikhov T.P. « Development of alternative energy in Uzbekistan is a question not technical, but economic one», Journal of « Economic Review », November, 2004.
9. Zakhidov R.A. Renewable energy sources is a new turn in power », Geliotekhnika, №2, 2002.
10. Energy Balances of Non-OECD Countries 2000-2001, IEA Statistics, 2003 Edition

Chapter 10

CURRENT STATE AND PROSPECTS OF RENEWABLE ENERGY TECHNOLOGY IN RUSSIA

Sergey Molodtsov

Centre for Energy Policy (Russia)

1. RESOURCE BASE OF RES

Along with vast proven recoverable fossil fuels reserves and about 90% of technical capability of hydro power currently not in use, Russia owns a unique energy potential of RES (Table 1).

Table 1 Energy potential of RES in Russia, mln tce/year

Source	Gross potential	Technical potential	Economic potential
Small Hydro	360.4	124.6	65.2
Geothermal Energy			115
Biomass Energy	10000	53	35
Wind Energy	26000	2000	10
Solar Energy	2300000	2300	12.5
Low Potential Heat	525	115	36
Total	2340000	4592.6	273.7

Source: IEA, 2003

A. Iacomelli (ed.), Renewable Energies for Central Asia Countries: Economic, Environmental and Social Impacts, 123–129.

Speaking about statistical data presented above one has to underline that economic potential of RES is a part of technical potential, which can be realized at the present level of prices for fossil fuels, heat, electricity and other externalities (equipment and materials, transportation and wages). Economic potential of RES (273.7 mln tce/year) given in the table 1 was evaluated in 1993. It is clear that everything is different now. Taking into account constantly growing cost of traditional energy sources and externalities this part of potential of RES in Russia should be much higher.

Therefore, technical potential, taking into account relatively high inertion of technological development, is more informative, especially for the horizon of 40-50 years. The estimated technical potential of RES is in 5 times higher than total primary energy consumption in Russia in 2002.

The availability of RES in each particular region of Russia (like any other country) depends on the landscape, natural and climate factors, in particular on wind speed, intensity of solar radiation, quantity of small rivers, etc. That is why some Russian regions are more attractive for deployment of RES-based technologies. The list of these regions related to each particular kind of RES is given below (Table 2).

Table 2 Regions attractive for wide-scale RES utilization

RES	REGIONS
Wind Energy	Arkhangelsk, Astrakhan, Volgograd, Kaliningrad, Maga-dan, Novosibirsk, Perm, Rostov, Krasnodar, Khabarovsk, Dagestan, Kalmykia, Karelia.
Solar Energy	The North Caucasus, the Black and Caspian Sea regions, Southern Siberia, the Far East, some regions in the South-East.
Biomass	Murmansk, Arkhangelsk, Karelia, Vologda, Komi, Pskov, Novgorod, Leningrad Oblast.
Geothermal Energy	Kamchatka, the Kuril Islands, The North Caucasus.
Small Hydro	The Norht Caucasus, the Urals, Eastern Siberia.

Source: P. Bezroukikh, 2004

2. CURRENT STATE OF RES UTILIZATION

Despite of huge potential of RES, their current use in Russia is very limited. The share of combustible renewables and wastes (crw) in total primary energy supply and electricity generation in the country is about 1-1.5% (Figure 1).

Figure 1. Share of RES in TPES and Electricity Generation

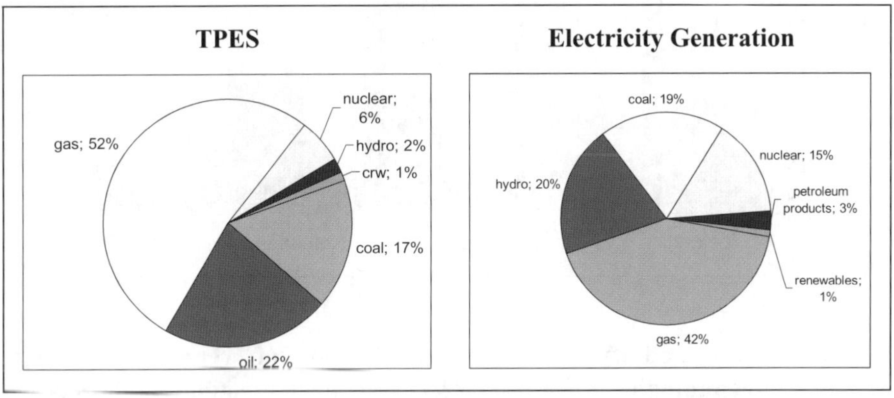

It is much less than in the leading industrially developed countries, dealing with wide-scale deployment of renewable energy technologies (Figure 2). There are several main factors influencing on limited use of renewables in Russia. The following ones are to be emphasized:

- **Consumers' preference to traditional, conventional technologies of energy production**. There are several reasons of this consumer's behavior. First of all it is necessary to say about existing lack of information and awareness of decision makers, potential consumers and general public on renewable energy technologies available at internal and international markets, their technical and economic characteristics. One more reason is limited governmental support (both financial and legislative) of RES development. In its turn it is a negative signal for private sector.

Figure 2. Share of Renewables in Electricity Generation in Russia and Foreign Countries

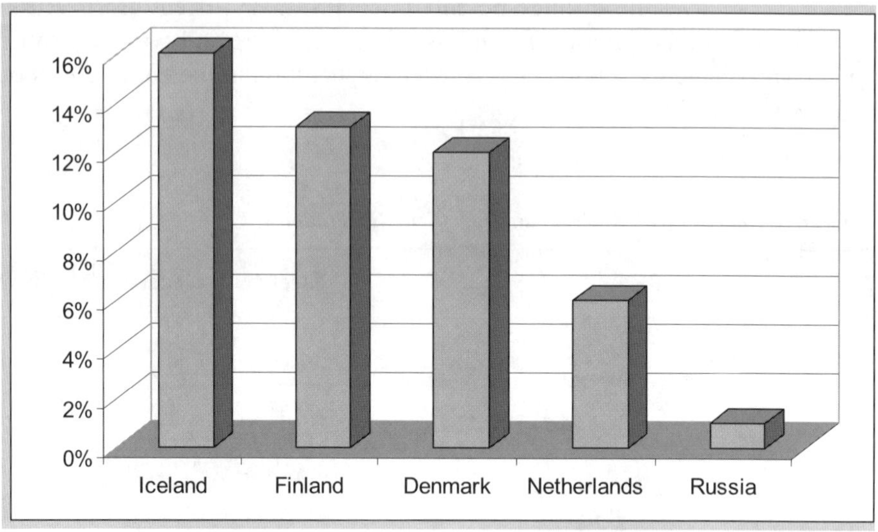

- **Low activity in commercialization of renewable energy technologies**. Starting R&D aimed at creation of RES-based equipment or technology, foreign (western) companies keep in mind that final destination of their efforts is commercialization (market promotion) of results achieved within implemented scientific and research works. In Russia there is a huge time and investment gap between R&D and market prospects of their results. There is a number of very promising renewable technologies developed by domestic scientists. Some of them have already achieved demonstration stage.

  Russian enterprises have the engineering and technical skill sufficient for wide-scale production of renewable energy equipment. Today, there are 100-150 Russian enterprises which can manufacture small and large scale RE systems, in particular:
  - wind energy systems from 0.04 to 16 kW;
  - water pumping wind energy systems;
  - autonomous solar PV-systems of capacity from 0.06 to 1 kW;
  - solar collectors and water heating solar systems;
  - micro hydro electric stations of capacity from 4 to 100 kW;
  - small hydro electric stations;
  - individual biogas modules;
  - heat pumps.

However, very few of them are commercially active, because there is no market demand for these systems in the country.

- **Low economic competitiveness.** One of the most serious obstacles to the market penetration of renewables is the high specific capital cost. One unit of installed capacity based on renewable energy technology costs several times higher than the same unit at coal or gas power plants. That's why electricity generated by renewables can not compete with electricity produced by power plants based on fossil fuels.
  Nevertheless, even now renewable energy projects could be cost effective and economically competitive in many Russian regions. First of all, it concerns regions with favorable natural conditions for renewable energy technology deployment and high net energy import dependency. These regions spend more than a half of their territory budgets on imported fuels. High transportation tariffs make imported fuels extremely expensive and create a serious social problem of energy affordability for the poorest part of population. It's also worth mentioning that due to relatively low installed capacity of electricity generating equipment renewable energy projects have lower capital intensity than, for instance, fossil fuel-based projects.

In order to make renewable energy technology competitive throughout the country large scale investments in R&D (as well as in commercialization activity) are needed. Leading industrially developed countries have a large governmental R&D budgets for renewables. For example, in 2002, United States invested in R&D in RES about 250 mln USD, Japan – about 130 mln USD. Unfortunately, Russia can not boast of comparable renewables R&D budget. It's very important to add that foreign private companies are also very active and that large scale investors of this sector of research, often invest much more than governments. In Russia, private investors practically ignore renewable energy sources development.

3. HOW TO OVERCOME EXISTING BARRIERS

Table 3 lists policy measures aimed at overcoming barriers which impede expansion of the scale of development of RES in Russia.

Table 3 Barriers which impede expansion of the scale of development of renewable energy sources in Russia and ways for overcoming

BARRIER	WAYS FOR OVERCOMING
Consumers' preference	Investment tax incentives, favorable loans; Accelerated depreciation of renewable energy equipment; Non-discriminatory access to the grid for independent energy producers using renewable energy equipment; Legislative measures make it necessary to install renewable energy equipment (in areas where favorable environmental conditions exist); Information and education campaigns; Demonstration project implementation
Low activity in commercialization	Creating state body responsible for commercialization of renewable energy technologies; Application of advanced foreign experience
Low economic competitiveness	Financial support of the state for R&D and demonstration projects in the field of RES; Attraction of the resources of international financial institutions, funds and programs; Introducing additional environmental payments and taxes for using fossil fuels; Reducing subsidies to the branches of conventional energy; transferring to the users the burden of full costs in terms of prices for fossil fuel, including the expenditures for environmental protection measures; Competitive price guarantee for independent energy producers using RES

A set of measures aimed at overcoming consumers' preference to conventional sources of energy includes financial, legislative and information measures, in particular, providing legislative (non-discriminatory access to the grid, obligatory installation of renewable energy equipment), and information support, including awareness on technical and economic characteristics of renewable energy installations available in the market and demonstration of their advantages.

The same measures could be useful in the field of commercialization of RES. It is also very important to create a state body responsible for renewable energy development in the country and for commercialization of RES in particular. Some foreign countries have already achieved definite progress in the market penetration of RES (Figure 2). Their experience could and should be applied in Russia. As far as low economic competitiveness of RES is concerned it could be increased through implementation of measures of direct financial support and measures creating favorable conditions for market penetration of renewables.

One more very important measure related to overcoming practically each existing barrier preventing market penetration of renewables is to be named. This measure is improving the international cooperation of Russia with foreign countries in the field of renewable energy development. It is necessary to develop cooperation with CIS countries, countries of Western Europe and other countries dealing with renewable energy technology deployment. This cooperation could be realized through:

- Scientific research based on bilateral and multilateral agreements;
- Joint implementation of demonstration and commercial projects;
- Organization of international seminars and educational programs on actual problems of current state and future development of RES;
- Intensification of activities of existing international organizations that are engaged in the development and advancement of renewable energy technologies and creation of new organizations of this kind that are capable of exerting a great influence to ensure rapid development of positive trends in setting issues on the greater utilization of RES.

REFERENCES

1. P. Bezroukikh. Present Day Situation and Prospects of Renewable Energy Sources in Russia. Energy Policy. Vol. 1. Moscow. 2004, pp. 3-19.
2. IEA. Renewables in Russia: From Opportunity to Reality. Paris. 2003.
3. IEA. Energy Policies of IEA Countries. Paris 2003.
4. IEA. World Energy Outlook 2004. Paris 2004

Chapter 11

THE HYBRID SOLAR – WIND SOURCE OF THE ELECTRO ENERGY AND PROSPECT OF ITS APPLICATION

R.I. Isaev, D.A. Abdullaev

Scientific Engineering and Marketing Research Center of the Communication and Information Agency of Uzbekistan

1. ABOUT THE EFFECTIVE UTILIZATION OF RENEWABLE ENERGY

A great part of technical objects and small settlements located in significant deserted and mountain territories, removed from an industrial electric network, are not provided with electro supply sources. The use of the renewable energy sources (RE) of solar radiation and air flows in the conditions of Uzbekistan can promote the decision of this important technical and social problem. On making the decision on the use RE on the removed objects it is necessary to study and take into account the following factors:

- The maximal capacity consumed by single or group of removed consumers;
- Meteorological conditions of district;
- The intensity of solar radiation, hourly average speed of the wind and the duration of these parameters during the day and months which define the scheme of utilization of RE, depending on requirements of the consumer;
- The operating mode of the consumer of the electric power - sporadic or continuous.

A. Iacomelli (ed.), Renewable Energies for Central Asia Countries: Economic, Environmental and Social Impacts, 131–139.

Among these factors the operating mode of the consumer is the most essential. So, the sporadic mode is typical to the equipment of RE-transformation, for the household requirements of illumination, water purification, and to the using by information means. In such conditions transforming of the one kind of RE is enough, for example, the energy of the sun or the wind with storage battery (SB) of the corresponding capacity.

To the operative working belongs the removed objects of telecommunication (the regenerators of the optical signals, base stations of mobile communication, radio and TV – re-transmitters), the meteorological posts, special objects of the control and safety concern. Even the short-term break of the electro supply to these objects concerned with the large fine sanctions and other technical and economic troubles: the use of the diesel generators is connected with the big material expenses (expenses on obtaining and on transportation of fuel), technical difficulties of service, necessity of duplication by reserve energy sources. As not less serious factor the problem of environmental contamination acts.

The territory of Uzbekistan is provided with high intensity of solar radiation (700-1000 W/m^2) in current of 2600-3000 hours in a year. There are significant local territories where average annual speed of the wind is 4-4,5m/s and more. In these territories the use of the hybrid solar - wind systems (HSWS) in amount of the source of the electric power is capable to solve the problem of sustainable energy supply of operative working objects, reducing up to the minimum of capacity of storage batteries and the troubles, having place at constant use of diesel generators polluting an environment.

2. ENERGY RESOURCES OF SOLAR RADIATION

Territory of the Republic of Uzbekistan belongs to the category of countries with high solar radiation intensity. Gross potential of solar energy is $50973 \cdot 10^6$ toe. Technically realizing potential is $176,8 \cdot 10^6$ toes. Unfortunately, this powerful energy source is still non-realized.

For effective transformation of solar energy into electric or thermal one, first of all it is necessary to know the solar elevation in different periods of year on given territory. On the most northern territories of Uzbekistan (45°35' n.l.) the most height of Sun in summer solstice reaches up to 68°, and on the most southern territories (37°10') is 76°. In winter solstice is 21and 29°, accordingly [1].

Data of table 1 shows the high energy potential of solar radiation on the whole territory of Uzbekistan and the necessity of regulation of PV-arrays inclination angle against the angle of sunrays incidence.

Considering data on winter and summer seasons we can conclude that in conditions of Uzbekistan solar power engineering with correct projecting can cover load demand up to 1/3 of summer and up to 1/6 in winter. Therefore, utilization of solar energy only is effective for energy supply to small settlements in desert and mountain territories.

Energy supply to technical objects of permanent operation in periods of absence of solar radiation is realized by use of additional energy sources. This circumstance caused the necessity to use other kinds of renewable energies, in particular, wind energy.

Table 1 Solar energy characteristics of Uzbekistan

	Characteristics	Winter	Summer
1	Solar elevation - north: 45° 35° n.l. - south: 37° 10° n.l.	 68° 76°	 21° 29°
2	Daily sunshine duration, hours/day	3 – 5 in average	10 – 13 8 - 10
3	Days without sun, days/year - north - south	 45-50 22-25	 10-15 1-4
4	Average sunshine duration, hours/year - north - south	 2800 3050	
5	Direct solar radiation intensity (S), kW/m² - on plains - on high-altitude stations - on Kyzylcha station, Smax	 0,80-94 0,94-1,06 1,21	

3. SHORTLY ABOUT WIND ENERGY RESOURCES

The gross potential of wind resources was defined 2,2-10^6 toe, technical potential -0,4-10^6 toe, [I]. Average yearly wind speed on the whole territory of Uzbekistan is 2-2,5 m/sec. This circumstance caused the opinion on non-prospectiveness of wind power engineering in the country. At the same time, research of many years conducted by CASRHI (Central Asia Scientific Research Hydrometheorological Institute) specialists showed that monthly, seasonal and yearly average wind speeds on the territory of the Republic are different. For example in Aral Sea basin, Kyzylkum desert territory and foothill zones of Tashkent region (Charvaq and Bekabad) yearly average wind speed is 4-5 m/sec and more [2]. Taking into account that these territories are remote to considerable distance from grid and on hard-to-reach desert, mountain and foothill regions, utilization of wind energy in such regions looks economically expedient.

Average potential capacity of wind flow, on the most part of the territory of Uzbekistan is 50-150 W/m^2. On the north, in Aral Sea basin and foothill zone of Tashkent region, it exceeds 150-200 W/m^2 [2].

Analyzing the given short information about wind energy cadastre of Uzbekistan we can conclude that there are large local territories with practically efficient average wind speeds. To utilize wind energy resources of such zones in 2003 SEMRC and CASRHI have started detailed study of wind regimes in a certain number of local zones. Results of these researches will allow making a correct estimation of the wind energy resources and will create a base for wide scale use of wind energy in Uzbekistan.

3.1 Hybrid solar-wind system for TV-transmitter Charvak, Uzbekistan

Complex of HSWS equipment was constructed and put into experimental operation in 2000 under financial support of European Commission in the frame of INCO-Copernicus Program (grant ICOP-DEMO-4068-98). Executors of demonstration project were SEMRC (scientific coordinator and principal investigator), firm Armines (project coordinator, France) with the participation of specialists from Portugal (firm F.F.Lda) and Uzbekistan (Institute of Power Engineering and Automation, firm Bakht-Ener).

HWSW and exploitation results

HWSW consists of the following:

- Arrays of PV-modules (Total Energy, France) of 60 m^2 total square with 6 kW peak capacity;
- 3 kW wind turbine (Southwest Windpower Co., USA);
- 1525 Ah/48 V storage battery (Oldham, France);
- measuring and controlling equipment;
- 4.5 kVA bilateral inventor.

In normal conditions when wind speed is over 3 m/sec or sky is clear or there are both factors together, energy produced by PV-arrays and wind turbine goes to consumer and at the same time charges the storage battery (SB). If charge of SB is full control system turns off the chain of damp load resistance and excessive energy is transformed into thermal one.

If both sources are absent (grey day or night and also there is no wind) customer is provided by energy during the time *At* by SB discharge. When SB is discharged till maximal allowed level, discharge controller is activated and reserve electric grid is connected. Electricity supply to the load and SB charging are made by grid. When wind or solar radiation are enough to make consumer operation occur, external chain is disconnected [3].

Measuring, scaling and recording the values of capacity production by PV-arrays, WT and whole HSWS, temperatures, maximal wind speed and number of other parameters characterizing modes of system operation, is made by Enerpak block. Results of more than 30 measurements are fixed in special tables.

3.2 Analysis of operation of hybrid solar-wind system

Average monthly meteorological data of Charvak site given by CASRHI were used to calculate a theoretical productivity of HSWS *(Ec)*. Curves characterizing total productivity of PV-arrays and HSWS are given on the fig.l and fig.2 accordingly. Here *Енс, Енг* are calculated and real production of electric energy by whole HSWS, *EL* is energy consumed by the load, *AE* is energy transferred to the external load. Analyzing these curves we can notice that:

- Real productivity of PV-arrays *Er* in summer period essentially less than calculated *Ec;*
- Average monthly and yearly real production of electric energy {Енг-12280 kWh) are less than calculated (EHc =1198 kWh and 14375 kWh). These facts are showing the decreasing of PV-arrays productivity in summer hot periods;
- Energy produced by HSWS generally covers the load demand during a year. At the same time, in winter periods of high cloudiness and absence of wind during 2-3 days, sustainable operation of TV-transmitter is provided by external source not exceeding l 000 kWh a year;
- Average monthly overproduction by HSWS of load demand is E -E 315 kWh monthly and 3830 kWh during a year. This energy can be transmitted to external consumer.

Therefore, the assumption on high effectivity of combined use of solar and wind energy has been confirmed, especially for energy supply of remote sites.

3.3 Analysis of stability operation of PV-arrays

Stability of PV-arrays in hot climate conditions on the basis of 3 years real exploitation data is discussed.

The energy production curves of HSWS *Ec, Er, Eav = f(T)* are demonstrated on fig. 3 Analysis of these characteristics shows that during 3 years of HSWS operation:

a) Value of E *=f(T)* in November-March differs from *Ee ~ f(T)* not essentially. This fact confirms that in considerably low temperature conditions, transformation coefficient of PV-array is up to 10%. The wind turbine in the good wind conditions covers the less productivity of PV-arrays of HSWS in this period;

b) In April-October *Er = f(T)* gradually decreases reaching the minimum in summer period. The temperature increase reaching 60°C on the PV-arrays' surface in June-August causes essential lowering the transformation coefficient of PV-array;

c) Also gradual decrease of *Er = f(T)* year by year is observed. This fact can be explained by changing of meteo-factors and probably by worsening the quality of PV-arrays. This phenomenon should be studied during a long time to estimate reliability and lasting qualities of PV-arrays. As a result of stated above factors, in hot period up to 1/3 part of solar energy potential is not utilized.

To estimate the level of temperature influence to PV-arrays productivity the indicator *Ke = Er/Ee* is used. The curves *Kec, Ker Keav ='f(T)* are given on fig. 4. Analysis of these curves shows that transformation coefficient of PV-arrays *Ker* significantly decreases in summer period, lowering up to 1/2 from *Kec.* This fact once more time confirms the known opinion, that the productivity of PV- arrays essentially depends on the temperature on its surface.

Stated facts show the necessity to develop the researches in such a direction to specify the influences of both temperature and atmosphere conditions (cloudiness, rain etc.) on decreasing the PV-arrays productivity. Therefore, creation of the variants of PV modules resistant to temperature changes should be an important direction of investigations and developments. Solution of this problem will allow more effective using of PV-systems in all countries with hot climate and long duration of sunny days.

4. APPLICATION OF THE HYBRID SOLAR-WIND SYSTEMS IN THE SPHERE OF COMMUNICATION AND INFORMATION OF THE REPUBLIC OF UZBEKISTAN

In the sphere of communication and information numerous objects requires the presence of guaranteed, reliable and highly stable sources of the electric power. It is connected with the use of the high technical and sensitive to non-standard fluctuations of the electric power of the informational – communication technologies in this sphere.

Power consumption of the electric power of the objects is various, from 3 up to 300 kW. Approximately the general power consumption is about 30 – 40MW.

The introduction of hybrid solar - wind systems should be carried out in three stages and in two operating modes:

- **The first stage** – the hybrid systems should be used in objects remote and removed from an industrial electric network. The example of such objects are the TV and radio-transmitters, located in the mountainous and deserted districts of more than 20 places in Tashkent, Samarkand, Navoi and other areas, the base stations of the companies of mobile telecommunication on the preliminary data more than 50 objects of companies Uzdunrobita, DaewooUnitel, Coscom and others. Power consumption of these objects changes in limits from 3 up to 15 kW. On separate objects achieves more than 100 kW. The hybrid systems can work mainly as the basic source of power supplies, and in some cases, as reserve.
- In **the second stage** – will be started the works on the use of hybrid systems on objects of the stationary type - in the post offices of communication and units of telecommunications with power consumption in limits from 5 up to 150 kW. Their approximately amount is about – 4000. In connection with the forthcoming program of large-scale introduction of electronic document circulation and electronic commerce are showed even more rigid requirements to objects of telecommunications, accordingly and to system of electro supply.
- **The third stage** - provides the use of the renewable sources of the electric power on objects where power consumption will exceed more than 150 kW. These are large units of telecommunications in the regional and regional centers, on cities, in objects special importance and in others. Above mentioned stages will promote

achievement of maintenance by reliable, stable and guaranteed power supplies of objects of telecommunications at economy of fuel and energy resources spent for development of the electric power of industrial purpose, the economy of diesel fuel spent by diesel generators used as reserve and emergency sources, and also reduction of emission of harmful gases in an atmosphere.

5. CONCLUSION

Experience of creation and exploitation of HSWS completely confirmed the high efficiency of combined use of solar and wind energies in the conditions of Uzbekistan.

Principles of combined use of renewable energies, results of HSWS exploitation were discussed at International Workshop "Hybrid-2002" organized by NATO grant EST-ARW-977881 (May 22-24, 2002, Tashkent). Workshop participants highly appreciated the results of made elaborations and investigations and recommended HSWS for wide scale implementation.

It is necessary to organize an investigations of diverse types of PV-cells and modules in real meteoconditions similar to Charvak sites ones, that allows estimating their

quality and stability and to recommend them for application in Central Asia region.

Authors note with the pleasure that this report is a result of generalized researches made in the frame of European Commission grant ICOP-DEMO-4068-98.

REFERENCES

1. Initial Communication of the Republic of Uzbekistan under the United Nations Framework Convention on Climate Change, Tashkent, 1999. p.p. 106

2. Rudak M.S. Wind-Helioenergy Cadastre of the Republic of Uzbekistan (in Russian), Tashkent, 2003, p.p. 147

3. Abdullaev D.A., Isaev R.I. Hybrid Solar-Wind System for Power Supply to Remote Sites, Proceedings WREC-1998, part IV, p.p.2697-2700

4. Abdullaev D., Isaev R., Mayer D. Operational Results of Hybrid Solar-Wind System in Charvaq, Uzbekistan, NATO International Workshop Hybrid-2002. 22-24 May, Tashkent, Uzbekistan, p. 5

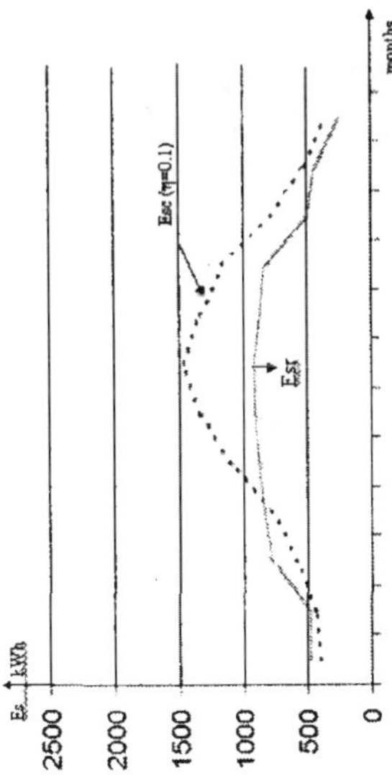

Fig. 1.

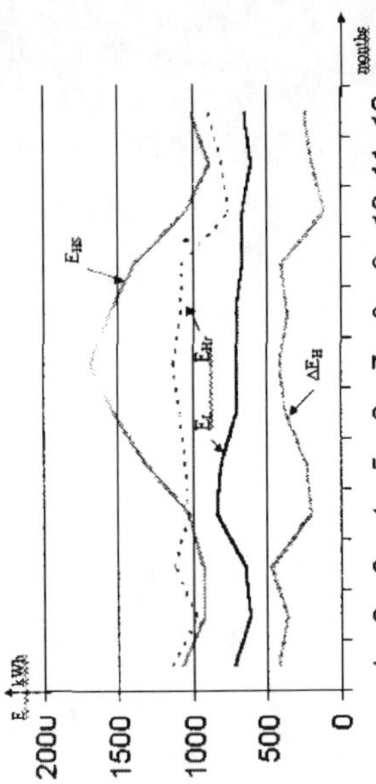

Fig. 2.

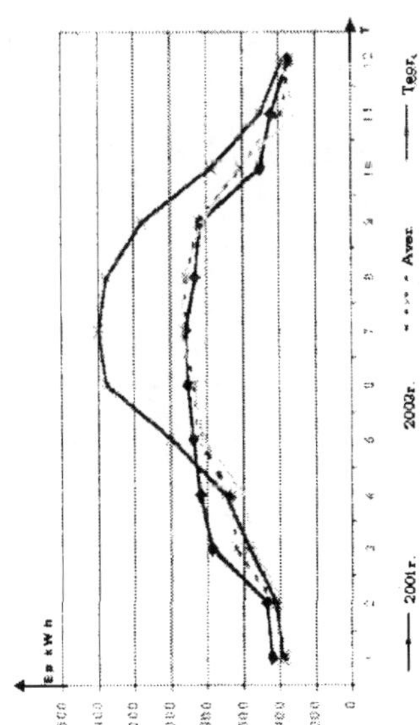

Fig. 3.

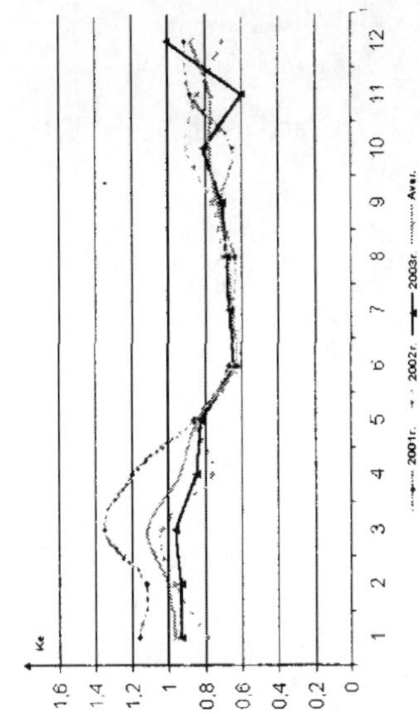

Fig. 4.

Chapter 12

NEW METHODS FOR IMPROVEMENT OF EFFICIENCY OF SOLAR CELLS ON THE BASIC Si-MONOCRYSTALS

R.A. Muminov, O.M. Tursunkulov

Physical-Technical Institute, Scientific Association "Physics-Sun", Academy of Sciences Republic of Uzbekistan, G.Mavlyanov Str.2B, Tashkent, 700084, Uzbekistan, *oybtm@physic.uzsci.net*

The alternative and renewable energy sources, like a wind power and a sunlight, hydro- and geothermal energy is of intense interest all over the world. And it is clear that limitation of traditional fuel-energy resources becomes obvious on the one hand and on the other hand the specified renewable energy sources mean ecological consideration. Among alternative and renewable energy sources a special place is occupied by direct transformation of sunlight to electric energy by means of semiconductor crystals. This method is the most convenient at the same time technological developed. The successes in the field of direct transformation of the solar energy in electric one by semiconductor crystals are determined by the creation of cheap and efficient technological ways of producing highly effective solar cells. Therefore there is a constant search of the new physical, technical and technological ways which studying should successful promote the production of effective semi-conductor photo-converters. In this connection crystalline silicon takes a special place. It is characterized by significant achievements in the scientifically substantiation essentially of new scientific, technical and technological approaches in perfection of production of solar cells on crystalline silicon. In particular we develop the new approaches to increasing of efficiency of solar cells (SC) on the basis of crystalline silicon.

A. Iacomelli (ed.), Renewable Energies for Central Asia Countries: Economic, Environmental and Social Impacts, 141–148.

1 THE METHOD OF DIFFUSION OF IMPURITY THROUGH THE OXIDE COVERING FOR FORMATION OF THE STRUCTURE OF SOLAR CELL.

Due to process of diffusion of the impurity in oxide covering this method allows to receive of the effective area of a the bulk charge in the structure of the solar cells. This method is in principle new in the world practice.

What feature consist of the given method?

It is known that it is necessary to formation shallow-lying p-n-junction for increase of the efficiency silicon SC. We developed the method of carrying out of process of diffusion of impurity through various oxide coverings for formation p-n-junction in mono-crystal silicon. In result the method has allowed to receive the positive features, in particular:

1) It is preliminary protection of the face surface of SC against erosion before formation p-n-junction.
2) At the formation of the face layer on a way of doping impurity the oxide covering provides a role of potential barrier. These barriers provide not only an opportunity of controlling junction depth and impurity concentration but also play a role of a filtration of foreign atoms.

   Beforehand of the process of receiving of diffusion of SiO_x oxide layer is resulted to reduction of diffusion depth of an impurity hence to improve of functional characteristics of SC (table 1).

Table 1.

The structure of SC n^+np^+	**Without oxide layer**	**Thickness of oxide layer 300-350 Å**	**Thickness of oxide layer 600-650 Å**	**Thickness of oxide layer 900-950 Å**	**Thickness of oxide layer more than 1000 Å**
The depth of p-n-junction	0,60±0,06	—	—	—	<0,35±0,03
Photoresponse at λ=0,4 micron	0,12 from maximum	0,15 from maximum	0,25 from maximum	0,31 от maximum	0,32 from maximum
Short-circuit current	920 mA	950 mA	980 mA	990 mA	1000 mA

2 CREATION ON BASE OF COMPOSIT MATERIALS OF ANTIREFLECTION COATINGS ON THE SURFACE SILICON SC ALSO IS IN PRINCIPLE NEW APPROACH AND PROVIDES HIGH EFFICIENCY FOR ABSORPTION OF THE MAXIMAL REGION OF THE SOLAR SPECTRUM IN BULK OF THE SOLAR CELLS AND INCREASE OF ITS POWER PERFORMANCE PARAMETER.

We solved such question by the creation on base composite materials antireflection coatings on the surface of SC silicon. The structure of silicon nano-supplement in SiO_x matrix as the specified coating were considered. It allowed to block all range of reflection coefficient and transmission for composite with film thickness 750 Å on basis Si/SiO_x with the various contents of silicon nano-supplement. The similar effect provided traditionally used film of nitride of silicon Si_3N_4 with thickness 750 Å. The spectral dependencies of reflection coefficient and transmission for the above mentioned coatings in silicon SC are shown in figure 1.

The other dependencies for silicon SC covered antireflection films are shown in figure 2,3.

Figure 1. The spectral dependence of reflection coefficient (a) and transmission (b) solar cell, covered with antireflection coating with thickness 750 Å: 1- Si_3N_4, 2-4 Si-SiO_2 with silicon containing; 2- 0; 3-0,25 (SiO); 4-0,28; 5-without antireflection coating

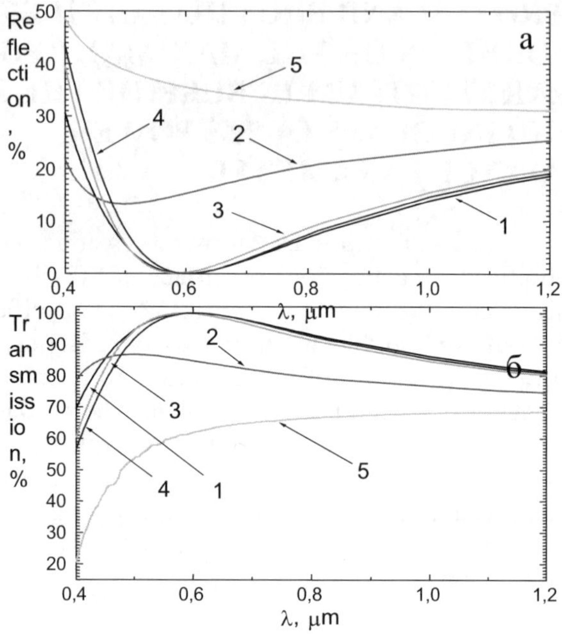

Figure 2. The spectral dependence ampere-watt sensitivity of solar cells covered with antireflection coatings with thickness 750 Å: 1- Si_3N_4, 2-4 Si-SiO_2 with silicon containing; 2- 0; 3-0,25 (SiO); 4-0,28; 5-without antireflection coating

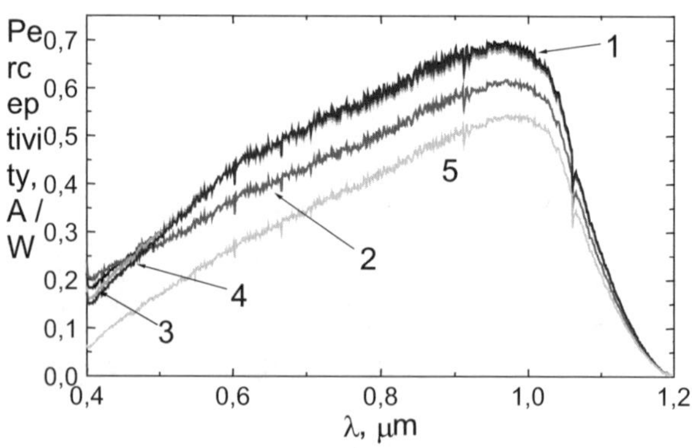

Figure 3. The dependence of density of output current of solar cell from composition of antireflection coating on the basis Si/SiO$_2$

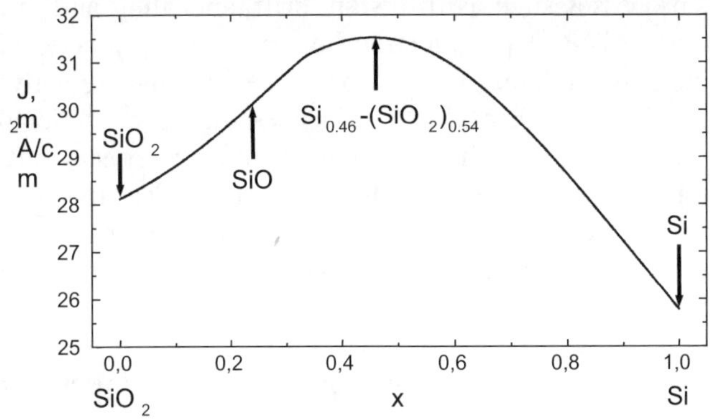

The analysis of the specified dependences shows, that composites on basis Si/SiO$_x$ allow to achieve characteristics of SC close to characteristics SC with coverings film of Si_3N_4. However the creation of composite coverings are ecologically favorably and technologically more effective.

3 THE THIRD METHOD DEVELOPMENT LOW TEMPERATURE TECHNOLOGICAL PROCESSES LIKE DIFFUSION, DRIFT, ANNEALING AND DEFECTS ELIMINATION OF THE CRYSTALLINE LATTICE IN CONDITIONS OF INFLUENCE OF ULTRASOUND WAVE TO A CRYSTAL ALSO HAVE NO ANALOGUES IN THE WORLD.

We made certain successes in this direction for last 10-15 years. The value of this approach is caused by the following reason: at high temperatures in the crystal bulk, surface and the metal - semiconductor border generate significant number of defects of a crystal lattice. It has strong influences on efficiency of the solar cells. At low temperature were carried out of the above-mentioned processes which provided full absence of the above-stated defects that considerably improves performance characteristics of solar cells.

Influence of the ultrasound wave (USW) on semiconductor crystals has three tendencies. The first includes preliminary influence of

USW field on semiconductor crystals with the purpose of ordering its initial defective structure and it has activation character. The second results to low temperature processes such as diffusion, drift, annealing and improvement of quality of a surface of p-n-junction border and p-n-junction itself. The third is acoustic-stimulated kinetic processes in the sensitive area of semiconductor devices when USW influences during operational mode of the device. The change of the energetic state of a crystal lattice of the semiconductor in USW influence is the general physical mechanism of the above-stated processes when as result of such influence there is a warming up of the charge carriers and phonons take place and also is observed their mutual moving.

The investigation of USW influence on the reverse current of saturation was carried out that the same SC was exposed in series to different processing.

We investigated acoustic-stimulated adhesive effect in semiconductor (figure 4) which means the changing of adhesive properties of metal films on semiconductor surfaces and increase in their adhesive strength after the ultrasonic treatment of the metal - semiconductor structures.

Fig. 4. The dependence of adhesive strength of the copper films on silicon substrate from the ultrasound treatment time. (sample Si-n №2,$N_{(O2)}$=4·10^{17} cm^{-3}, f=15 MHz, P=0,5 Vt/cm^2, T=293K)

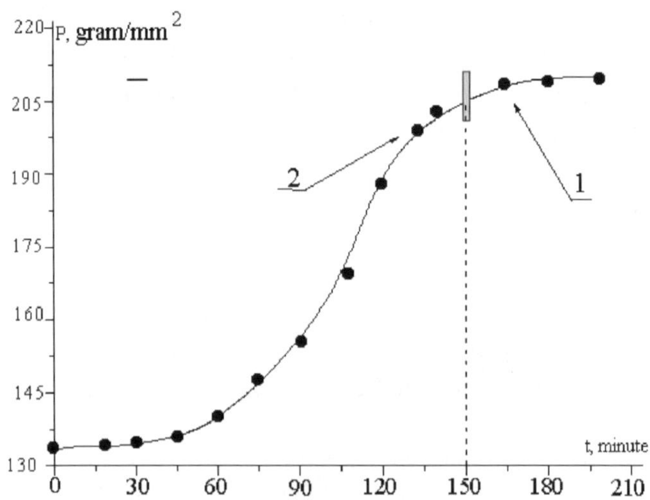

The investigation of the behavior of the reflection coefficient of optical irradiation R (λ) from a semiconductors surface subjected by ultrasound influence (figure 5). It is established that a degree of changing of R (λ) dependence determined by power, frequency and duration of USW influence. The acoustostimulation change of subsurface layer of the semiconductor crystal occurs with the diffusion impurity atoms from subsurface region of crystal in its depth, and also disintegration of some type congestion.

The efficiency of solar cells is defined by many factors and in particular the drift of carriers through a sample after generation of electron-hole pairs under sunlight. In this process for the value of signal amplitude is important the losses of carriers by traps. It is established that if the sample treat with USW then the dependence of signal amplitude on the drifting field will be vary from USW value.

Fig. 5. Spectral dependences of the reflection coefficient R with different degrees of doping of the n^+-Si layer of silicon measured before and after ultrasound treatment capacity $P_m \leq 1$ W/cm^2; frequency $f = 15$ MHz; Si-sample №1: $N = 1{,}75 \cdot 10^{20}$ cm^{-3}; before (curve 1) and after (curve 4); Si- sample № 2, $N_2 = 4{,}5 \cdot 10^{19}$ cm^{-3} ; before (curve 2) and after (curve 5); Si-sample № 3, $N_3 = 7{,}8\ 10^{18}$ cm^{-3} before (curve 3) and after (curve 6); $t > 60$ min

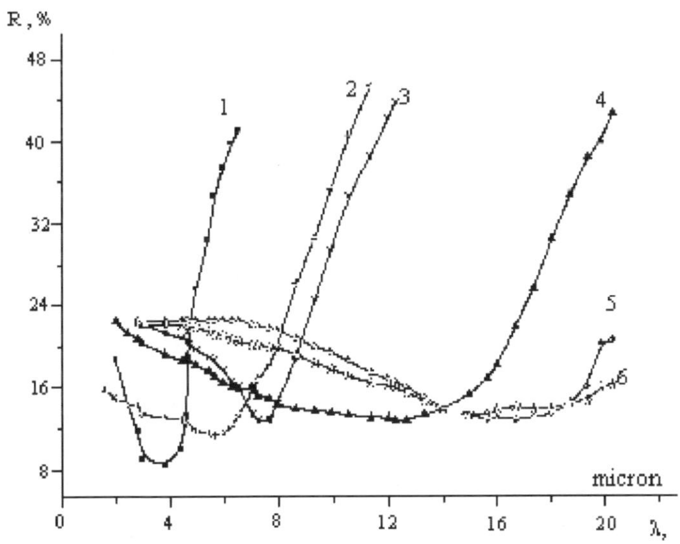

Thus the certain new step in the development and creation of highly effective solar cells on the basis of crystalline silicon is made in present paper. These achievements are directly applied in development of the real self-contained module power systems for example: the various type and purpose portable installation of electric illumination; self-contained photoelectric installation for power supply of incubators.

The certain achievement of our scientists, as a whole work, in development and investigation of renewable energy sources is presented in this paper.

Chapter 13

DESIGN OF SEMICONDUCTOR NANOSTRUCTURES FOR SOLAR CELL APPLICATION

L. Nosova[1, 4], S. Gavrilov[1, 2], I. Sieber[3], A. Belaidi[1], L. Dloczik[1], Th. Dittrich[1], A.A. Saidov[4], P.K. Khabibullaev[4]

[1] Hahn-Meitner-Institute Berlin, Abt. SE 2, Glienicker Straße 100, D-14109 Berlin, Germany;
[2] Moscow Institute of Electronic Technology, Moscow, 124498, Russia;
[3] Hahn-Meitner-Institute Berlin, Abt. SE 1, Kekuléstr 5, 12489 Berlin, Germany;
[4] Heat Physics Department, Katartal str. 28, 700135 Tashkent, Usbekistan;
Corresponding e-mail: lyusya7@yahoo.com.

1. INTRODUCTION

Last decades one-dimensional materials have been a focused research field both because of their fundamental importance and the wide-ranging potential applications in various electronic and electrochemical systems to be used in solar cells, catalysts, chemical sensors and many others.[1-3] One of the factors driving current interest in nanotechnological research is the perceived need for the further miniaturization of both optical and electronic devices.[4] There are many experimental approaches to fabricate nanowires, utilizing a variety of nanofabrication techniques[5,6] and crystal growth methods.[7,8] Because the growth is controllable almost exclusively in the direction normal to the substrate surface, electrochemical synthesis in a template is taken as one of the most efficient methods in controlling the growth of nanowires and has been used to produce a variety of metal nanowire arrays.[9] However, studies on the preparation of semiconductor nanowires by electrodeposition using templates are still scarce.

A. Iacomelli (ed.), Renewable Energies for Central Asia Countries: Economic, Environmental and Social Impacts, 149–155.

Thus, we have used anodic alumina as a template material. Nanoporous anodic alumina has many favorable characteristics as a template material for nanoparticle array fabrication because the template formation process is very simple and results in high density of parallel pores. The pore diameter can be tuned from ~10 nm to several hundred nm by varying the anodization conditions. Due to cylindrical pore geometry and monodisperse diameters, corresponding cylindrical and well-distributed nanostructural materials are obtained.

Semiconductor nanowires are of considerable interest as they are capable of being used in optical devices and microelectronic technology.[10] Cadmium sulfide is a typical wide band gap II-VI semiconductor having a band gap of 2.42 eV at room temperature. It has many commercial or potential applications in light-emitting diodes, solar cells or other optoelectronic devices. The synthesis and study of cadmium sulfide nanowires should stimulate much research and technological applications. In this study we fabricated metallic (Cd, Cu) nanowires into porous anodic alumina by using electrochemical deposition and then converted them into their chalcogenides by using sulfurization in H_2S. However, the electrochemical deposition of these materials on an insulating and rather thick barrier layer at the pore tips is not straightforward. High potentials are required for the tunneling of the electrons through the barrier layer. Electrodeposition on the barrier layer by direct current is very unstable and uniform filling of the pores cannot be achieved. This is due to a cathodic side reaction, which leads to a partial removal of the barrier oxide, formation of holes in the barrier layer, and local deposition in these pores. We have developed the electro-prethinning of the barrier layer to make electrodeposition possible and effective. CdS and Cu_xS nanowires were obtained as a result of ac electrochemical deposition of Cd and Cu into pores of anodic alumina and their subsequent sulfurization. CdS/CuS heterostructure based on porous alumina is proposed for solar cell application.

2. EXPERIMENTAL SECTION

Pure (99,99%) aluminum foils were rinsed in acetone and annealed in vacuum at 400°C for 1 hour in order to recrystalize the samples, enhance the grain size in the metal and therefore to get homogeneous conditions for pore growth on larger areas. Then aluminum foils were polished chemically by dipping them at first to the aqueous solution of NaOH (100g/l) for 0,5-1 min. at 50-60°C, short rinsing with distilled water and then dipping to the aqueous solution of HNO_3 (350-450 g/l) for 0,25-0,5 min. at 18-25°C, and then Al foils were washed with distilled water again.

Porous alumina films were formed by using two-step anodization process.[11] The anodization was performed under galvanostatic conditions 10mA/cm^2 at room temperature with vigorous stirring. Three kinds of electrolyte 1.2M sulfuric, 0.3M oxalic, and 0.3M phosphoric acid solutions were used to obtain the pore diameter of 10, 60, and 120 nm, respectively.

The first oxide layer was dissolved in a solution of 6 wt % H3PO4 and 1.8 wt % CrO3 at 60°C. The second anodization was performed under the same conditions as the first one.

There is always dense insulating so called barrier layer between porous part of alumina and Al substrate. It creates difficulties for charge transfer in electrolyte/substrate system during electrochemical deposition into pores. The thickness of the barrier layer strongly depends on voltage applied during anodization process. The porous alumina grown in phosphoric acid solution has the thickest barrier layer (about 130 nm thick), for porous alumina obtained in oxalic acid electrolyte the barrier layer thickness is about 65 nm, in sulfuric acid solution – about 10 nm. To reduce the thickness of the barrier layer we have used the following procedure. When the anodization was complete, the current density was stepwise decreased (10, 5, 2.5, 1.2, 0.6 mA/cm^2) over a period of 5 min while anodization voltage falls till 4-5V. This procedure leads to reduction of the barrier layer until 5-7 nm. At the end of anodization and prethinning processes porous anodic alumina films kept on aluminum substrate with quite uniform, parallel, and opened only at oxide electrolyte interface pores are obtained. These regimented nanoporous alumina templates were used for ac electrochemical fabrication of the CdS and Cu_xS nanowires.

The electrochemical deposition of Cd and Cu was performed in a pulsed voltage regime with rectangular pulses. Cu nanowires were fabricated by pulsed electrodeposition from a mixture of 35 g/l $CuSO_4$ and 20 g/l $MgSO_4$ aqueous solution, pH=1,3 was reached by adding of sulfuric acid solution. For Cd nanowires an aqueous solution of $CdSO_4$ was used as a source of Cd ions. Cathodic potential of -10 V was applied for 1 ms. Anodic waiting pulse (potential +3 V) was applied for 100 ms. The barrier layer is discharged during the anodic waiting pulse and metal cations diffuse to the electrolyte/electrode interface. Electrochemical deposition was performed until the metal reached the sample surface. After electrochemical deposition, the samples were rinsed in distilled water and dried in N_2-stream. The samples were sulfurized in a quartz tube furnace in Ar/H_2S (5%)-atmosphere (pressure 0.1 MPa) at 500° C for 3 h.

The cross sections of the samples were investigated by scanning electron microscopy (SEM, JEOL 4100). Phases were analyzed by x-ray diffraction by using the Cu Kα line of a Bruker Axs D8 advanced diffractometer. Photovoltage spectra were measured in the capacitive arrangement. The excitation was performed with a Xe-lamp and a quartz-prism monochromator. The photovoltage was measured as the signal in phase and phase shifted to the modulated light (chopping frequency 16 Hz) as well as the amplitude.

3. DISCUSSION

Figures 1a and 1b present SEM images showing the cross-section of PAA/Al after 2 step anodization process in 0,3M oxalic acid solution. Pore diameter is about 60 nm, the interpore distance is about 100 nm, and the thickness of the barrier layer is nearly equal to the wall size. SEM image (1a) shows that pores form well-defined hexagonal honey comb structure. The barrier layer after prethinning procedure is presented in Fig.1b. During prethinning procedure a lot of narrow pores are formed and the barrier layer becomes like tree-root net.

Figure 1. SEM micrographs of cross section of porous anodic alumina grown in 0,3M oxalic acid solution top part (a); bottom part after prethinning of the barrier layer (b); CdS (c) and Cu_xS (d) nanowires grown into porous alumina matrix.

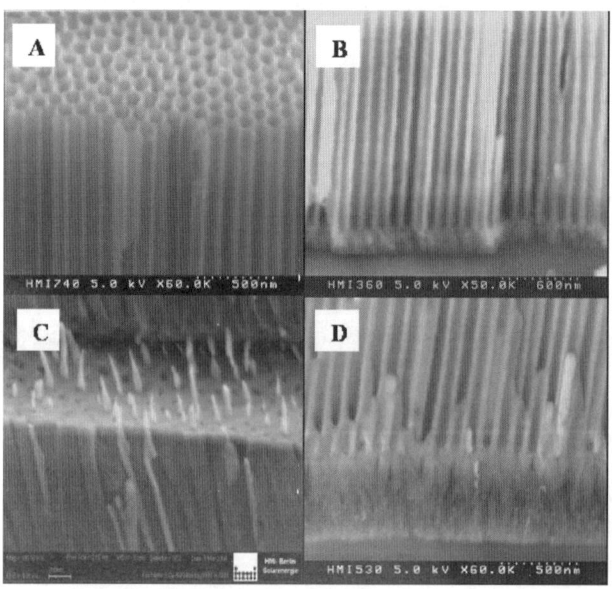

The two-step anodization method was used to make the porous structures more regular and uniform. During the first anodization step a self-ordering takes place. It was suggested that the repulsive forces between neighboring pores caused by mechanical stress at the metal/oxide interface promote the formation of hexagonally ordered pore arrangements. After removing of the porous alumina layer formed in the first anodization the information about self-ordering still occurs on Al surface. Thus, anodization of the remaining aluminum layer under the same conditions resulted in anodic aluminum oxide nonporous arrays of better uniformity and straighter pore channels. Pores grow perpendicularly to the Al substrate (Fig.1b). Straight pores are desirable for the electrodeposition of materials at the bottom of every pore. Pretreatment, such as annealing and electropolishing, plays an important role to obtain such well-ordered nanopore array.[12, 13]

After Cu deposition into pores the barrier layer is not observed any more, Cu is penetrated everywhere in PAA/Al interface (Fig. 1d). Similar results are obtained on ac electrochemical deposition of Cd into pores of PAA (Fig.1c). After sulfurization CuxS and CdS nanowires were fabricated. In the SEM images the nanowires are often interrupted and many pores are empty. It causes due to the cracking procedure of the sample which is needed for cross section preparation.

Figure 2 shows the spectrum of the photovoltage amplitude of the CdS/PAA/Al sample. The photovoltage amplitude starts to increase at about 1 eV and a second sharp increase appears at about 2.4 eV. The sharp increase at about 2.4 eV corresponds to the onset of optical absorption near the optical band gap. The presence of the low-energy photovoltage signal in the range from 1.0 to 2.4 eV is caused by a high concentration of defect states in the band gap. The origin of these states is not clear yet. To our opinion, mechanical stresses should play an important role for the formation of defects.

Figure 2. Photovoltage spectra of the CdS/porous alumina structure.

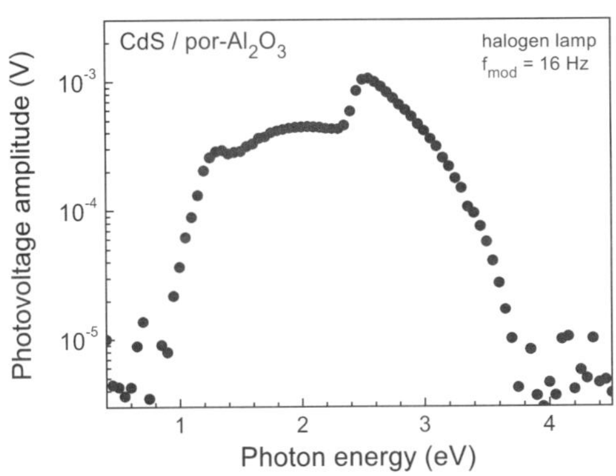

We propose to use porous alumina matrix kept on Al substrate as a template for fabrication of semiconductor nanostructures with p-n junction for solar cell application. The scheme presented on Fig.3 includes several processes: preparation of porous alumina with prethinned barrier layer, deposition of metal (for example, Cu), dissolving of alumina (it could be done in 1M NaOH at 25 °C) to get free-standing metallic nanowires on al substrate. Then one can make electroplating of the same metal to cover the aluminum. After that the metallic nanowires could be sulfurized by annealing in H_2S, as it was described previously, to convert the metal into its chalcogenides (in our case, into Cu_xS). CdS could be directly deposited onto CuS surface or deposition of metallic Cd with its subsequent sulfurization could be performed to obtain the heterostructure with p-n junction.

Figure 3. Scheme of solar cell fabrication

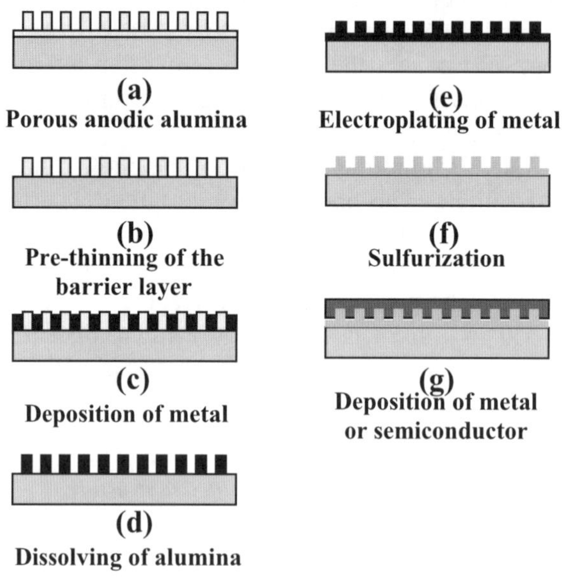

4. CONCLUSION

In our study the highly ordered porous alumina structures with pore size of 10 until 150 nm were fabricated. The diameter of the nanowires can be controlled by the pore diameters of the template. An advantage of proposed technology is that porous aluminum oxide layer is kept on Al substrate and becomes more suitable for the fabrication of photoelectrical nano-devices. Porous anodic alumina serves as a nice template for nanowires fabrication and investigation of their physical properties.

Developed approach gives opportunity to deposit a number of other materials: metals, their oxides, sulfides, chalcogenides, and hydroxides, into porous alumina matrix. The suggested procedure of nanostructure fabrication is relatively cheap, efficient, well-controlled, and easy for realization.

5. ACKNOWLEDGEMENT

L.N. is grateful to Duetscher Academischer Austauschdienst and Fundamental research Foundation of the Science and Technology of the Republic of Uzbekistan (grant FFI-2.1.41) for financial support.

REFERENCES

1. Xu, D. S.; Guo, G. L.; Gui, L. L.; Tang, Y. Q.; Shi, Z. J.; Jin, Z. X.; Gu, Z. N.; Liu, W. M.; Li, X. L.; Zhang, G. H. *Appl. Phys. Lett.* 1999, 75, 481.
2. Chen L., A.J. Yin, J.S. Im, A.V. Nurmikko, J.M. Xu, J. Han,. *Physica Status Solidi A* 2001, 188, 135.
3. Mozalev A., S. Magaino, H. Imai, *Electrochim. Acta* 2001, 46, 2825.
4. Stroscio, J. A.; Eigler, D. M. *Science* 1991, 254, 1319.
5. Ono, T.; Saitoh, H.; Esashi, M. *Appl. Phys. Lett.* 1997, 70, 1852.
6. Namatsu, H.; Horiguchi, S.; Nagase, M.; Kurihara, K. J. *Vac. Sci. Technol., B* 1997, 15, 1688.
7. Morales, A. M.; Lieber, C. M. *Science* 1998, 279, 208.
8. Martin, C. R. *Sciencc* 1994, 266, 1961.
9. Foss, C. A.; Gabor, Jr.; Hornyak, L.; Stockert, J. A.; Martin, C. R. *J. Phys. Chem.* 1994, 98, 2963.
10. Pandey, R. K.; Mishra, S.; Tiwari, S.; Sahu, P.; Chandra, B. P. *Solar Energy Mater. Solar Cells* 2000, 60, 59.
11. Masuda, H.; Fukuda, K. *Science* 1995, 268, 1466.
12. Li, A. P.;Mu¨ller, F.; Birner, A.; Nielsch, K.; Go¨sele, U. J. *Vac. Sci. Technol. A* 1999, 17, 1428.
13. Xu, T.; Zangari, G.; Metzger, R. M. *Nano Lett.* 2002, 2, 37.

Chapter 14

UTILIZATION POSSIBILITIES OF RENEWABLE SOURCES OF ENERGY IN SOUTHERN KAZAKHSTAN BY THE EXAMPLE OF KARATAUSKO-UGAMSKI ENERGY COMPLEX

T.K. Koishiyev

Head of Renewable Energy Sources Department of Kazakh State Univertisy

In the future renewable sources of energy (RSE) will play remarkable role in energy supply.

Even today RSE can play substantial role in power supply of decentralized consumers. This field of power engineering needs to be developed first of all in regions with abundant renewable resources and insufficient fossil fuel or absence of the latter.

Today utilization of solar, wind and hydroelectric power stations of modular type combined into single energy complex proved to be technically and economically more expedient.

Many foreign projects, which are intensively developed by specialists from USA, Japan, Italy, Spain, French and other countries, apply such combined schemes.

The basic purpose of pilot combined electric power stations operating on RSE is to supply electricity and provide opportunity for conduction of experimental researches and comparison of various technologies. Such opportunity allows accumulating of scientific data, which is vital for creation of more powerful and more effective combined electric power stations.

Favorable conditions in Kazakhstan allow utilization of RSE for the purpose of production of electricity and highly potential heat. Taking into account RSE when forming regional energy balance provide remarkable

A. Iacomelli (ed.), Renewable Energies for Central Asia Countries: Economic, Environmental and Social Impacts, 157–162.

reduction in consumption of traditional fossil fuel and substantially decrease pollution.

Harmonious combination of different kinds of RSE in general energy balance in Kazakhstan is the problem of primary importance.

In the present work questions related to construction of Karatausko-Ugamski energy complex of capacity of 170 MW operating on ecologically clean sources of energy of southern regions of Kazakhstan have been considered and the question, concerning development of general methods of modeling of operating mode of single subsystems of energy complex, has been formulated.

Construction of combined energy complex on the basis of RSE (solar, wind and hydro energy) in spurs of Karatau mountains near the Urgam river is the prototype of future complex utilization of RSE.

Following parts are included into composition of the complex:

1. Construction of two wind power stations (WPS) of capacity of 20 and 40MW;
2. Construction of complex of solar photoelectric stations (SPS) and solar collectors (SC) at the basic energy power stations aw well as at remote sparsely populated regions of total capacity of 10 MW;
3. Construction of complex of hydroelectric power stations on the Urgam river of capacity of 50 MW;
4. Construction of thermal electric power stations (TEPS) operating on gas with exhaust-heat boilers with capacity (3x17) MW in Turkistan.

Combined WPS and SPS –WPS-SPS are planned to be constructed at "Chayan" platform of capacity of 20 MW and at "Baizhansai" platform of capacity of 50 MW. SPS and SC with capacity 10 MW are also planned to be constructed.

Wind platforms for WPS are selected basing on the data concerning wind potential from weather stations. Average annual speed of wind on these platforms exceeds 5.5 m/s, which allows construction of two WPS.

On the territory of Kazakhstan the amount of solar energy that falls upon the surface per year in the form of solar radiation ranges from 4 to 7 thousands MJ/m^2 of intensity up to 700-800 W/m^2 depending on the region.

Duration of solar radiance on the territory of Kazakhstan is remarkably high (up to 3100 hours). At that the territory where solar radiance is observed during 2800-3100 hours is equal 1900,5 thousands of km^2 that makes up 70% of total area of the territory of Kazakhstan.

Consequently, long duration and substantial height of midday sun is typical to 2/3 of the territory of Kazakhstan.

Qualitative estimation of possible utilization of solar energy is determined by the ratio of the power of the solar station to the area under that installation.

Complex of hydro electric stations (HES) – on the Urgam river construction of hydro electric stations of total capacity of 50 MW is planned. The river basin is located on the territory of Kizgurtsk district of Southern Kazakhstan region.

Total effective volume of all reservoirs of cascade of Urgam River is equal to 136.3 mln. m^3 that is nearly 24% of average annual volume of the water flow of the river. This allows full regulation of seasonal flow of water. Specific water content is sufficiently high and up to six HPS can be constructed with total capacity of 50MW.

Turkestan TEPS – construction of three TEPS on the basis diesel-generators of total capacity of 17 MW equipped with exhaust-heat boilers is offered. To select suitable diesel-generator, which is the basic piece of equipment, several kinds of generators of various manufacturers were considered. In accordance with technical features generator manufactured by Finnish company "Vyartsilya" appeared to be the most suitable one.

Karatausko-Ugamski energy complex is a complicated and large scaled project, on different stages of its construction and exploitation many problems of theoretical and experimental nature might occur, which would require corresponding solutions.

The spectrum of intended researches is extremely wide because of the scale of the project and diversity of offered technical solutions concerning various techniques of energy transformation.

Successful accomplishment of the project requires deep theoretical investigations and creation of specific kind of mathematical models. Moreover a complex of experimental investigation is urgent.

We have developed general functional structure of multifactor mathematical model of operation of basic subsystems of Karatausko-Ugamski energy complex (KUEC), which operate on the basis of RSE .

It includes following models of operation of basic subsystems of the station:

I Model of formation of energy resources of RSE in the KUEC region;

1) Sun movement and solar resources model;
2) Wind resources formation model;
3) Hydro resources formation model;

II Functioning of solar photoelectric stations (SPS) and solar collectors (SC) model.
III Wind power station functioning model;
IV Hydro power station functioning model.
V Thermal electric power stations functioning model (diesel-generator);
VI Mathematical multifactor model of optimization of KUEC operation mode:
4) Model-project;
5) Model-experiment.

In general this kind of problem is rather difficult and a multifactor one. At present time there are no mathematical models which can solve such kind of problems in "corpore".

When developing mathematical models of single subsystems each of this models was divided into series of sub-models aimed to solve specific task.

Calculation algorithm is build in form of hierarchy of repeating cyclic subsystems inserted one into other. External cycles organize the sequence of searches of considered versions of calculation and can be altered depending on the task given at a certain stage of investigation.

The model of formation of energy resources of RSE in the KUEC region in general is the model designed to get primary information (total solar radiation, speed of wind and speed of water flows). Sun movement model fixes the sequence of determined relations describing the movement of the Sun in the sky. Sun movement model determines the most probable operation mode of the system and tracks dynamics of total solar radiation received by effective area of basic subsystems during working day.

In general "The model of formation of energy resources of RSE in the KUEC region" covers processing of climatic data of the region where stations are location. Primary information is processed and is displayed in the form which is convenient for calculation of energy resources of a given region.

Models of operation of solar, wind and hydro power stations analyze the influence of climatic, constructive and arranging parameters on the operation mode of single energy installations from point of view concerning reliability of functioning of the complex as whole.

Multifactor mathematical model of optimization of operation mode of KUEC (model-experiment and model-project) accomplishes selection of optimal and arranging solutions of single subsystems in accordance with system consistency of parameters of single modules of energy complex.

This large and difficult problem is solved by means of solving of questions such as creation of working model. Criterion function of the optimization problem is included in the values of average-exploitation power and effective thermal flow.

Multifactor mathematical model of KUEC will provide comprehensive investigation of energy and exploitation features and assess the efficiency of functioning of some single modules and the whole complex.

Preliminary estimations indicate that expenses required to accomplish the construction of KUEC of total capacity 170 MW are nearly 200mln. $ US and the production of power requires 8-10mln. $US. Production of electrical energy will be nearly 858mln.kWh and thermal energy 194350Gcal. Average annual prime cost of electrical energy during utilization period would not exceed 0.9 cents per 1kWh and the payback is estimated to be 5-6 years.

Developed agriculture and industry of the region and high population density allow quickly return of invested means.

Combination of various kinds of energy resources provides stable supply of electricity and thermal energy to consumers.

CONCLUSION

1. Realization of Karatausko-Ugamski energy complex of capacity of 170 MW is offered.
2. It is worth noting that given economical estimation is a preliminary one and bears conditional character since the dynamics of tariff and possible quota on releasing of greenhouse gases is not taken into account.
   Moreover, social aspects related to realization of the project are also not taken into account. However, such estimation indicates the possibility of accomplishment of the project under soft credit even with today's cost of electricity.
3. The work development and optimization KUEC needs to be continued and technical and economical assessment, which requires 40-60 mln. tenge (national currency) must be accomplished.
4. Direct benefits from realization of the project:
   - reduction of pollution;
   - development of additional sources of energy in energy deficient region;

- utilization of HPS and TEPS to cover demands of energy systems during rushhours;

5. Indirect benefits:

- reduction of CO_2;
- Solving of social questions and the problem of energetic independence of the Republic of Kazakhstan.

Chapter 15

RENEWABLE ENERGY: ENVIRONMENTAL AND NATURE PROTECTION ASPECTS

Kulsina Kachkynbaeva

CAREC Kyrgyz Branch Director

People from time immemorial are using energy as force of wind, falling stream of water, heat of the sun, burnt up organic fuel (wood, pressed dung, coal, gas, oil products) to satisfy permanently extending personal and social needs. The more developed social society had become, the more energy it needed in. The more advanced social society had become, the more knowledge and means people spent to find more powerful and effective kinds of energy.

Getting mechanical, thermal energy was continued by getting electric power. Electric power fundamentally changed people's overviews concerning possibilities of utilizing energy to satisfy their economic, social and cultural needs. If at the end of the XIX century electric power played, in general, an auxiliary and insignificant role in global balance. In 1930 about 300 billion watts - hours of the electric power were produced in the world. But in 2000 30 thousand billions watts - hours were already produced. Huge figures, huge rates of growth in presence of one or two generations of people!

And still, even with inventing of electric power, there weren't enough energy to satisfy people's needs. Together with inventing atomic energy, power rate of which was incomparably more then of those sources of energy used before, people did not only continue to open up the Earth, her ground, underground, water and air riches but they started to investigate the space.

A. Iacomelli (ed.), Renewable Energies for Central Asia Countries: Economic, Environmental and Social Impacts, 163–173.

And all the same there was no enough energy. The need in energy is growing faster, and the population is growing more and more. Huge power programs are working out, realization of which will demand enormous effort and great expenses.

It is obvious, that the level of material life, and, finally, of mental culture of people is in direct dependence on quantity of the energy available at their disposal.

Energy resources are any sources, which produce energy. They are classified according to their sources, dislocation, speed of use up, possibility to self-rehabilitation and other characteristics.

Primary power resources as a rule are divided into:

- **non-renewable** ones (coal, oil, slates, natural gas, uranium ore);
- **renewable** ones (wood, water-power, wind power, energy of the sun, geothermal energy, peat, biomass).

Usually people use energy sources available at their disposal by three ways.

1. First, they get thermal energy by burning ground and fossil fuel or catching energy of solar beams, and directly use this energy to heat various premises, including houses, and also to prepare food.
2. Second, they transform thermal energy contained in fuel or mechanical energy of falling water or wind into some work, for example, use it for starting to work various equipment, machines, cars, trains, etc.
3. Third, they transform mechanical, thermal energy and energy, liberated from nuclear disintegration, into electric power, and then they use it to produce heat or to do wide spectrum of works.

Any kind of energy resource is a part of the environment, surrounding a human being.

But when a human being for the first time seriously began to think over the phenomenon of interrelation and interdependence between energy and environment protection? May be when the forests began catastrophically to thin out, and then disappeared on the vast territory of the planet and that led to desertification? May be when struggle for owning oil and gas deposits in increasing frequency was resulted in destructive wars? May be when people, having possessed enormously powerful source of energy as atom, did not think over up to details of an opportunity to control it in case of running away as it was with the Chernobyl atomic power station?

In fact, it is impossible to get a simple answer to this question. For a long time the mankind was keen on utilizing energy to satisfy its growing needs, first of all in economic sphere, that very often it was luxury to think seriously over environmental protection. Even then, when on some continents the environmental problems were quite obvious, for example, mass ruthless cutting down of woods in Africa during many years without any reforestration activity, which led to desertification of a significant part of the continent, sharp reduction of the biodiversity and climate change, the mankind kept Olympic composure.

And only when environmental problems became so obvious and menacing in a global scale, when it was already impossible to hide from them because they began to threaten all animate nature on the Earth; when it became obvious, that burying in oblivion the principle «**Protect the Nature!**» as a boomerang reflected on a well-being of our planet and people living on it; only then people really seriously began to think over solving environmental issues, including power supply ones.

Beginning from the '80s of the twentieth century the world community started to declare openly that the environment protection issues were the issues of global concern priorities to achieve sustainable development. The new readout of time began from that period concerning interrelation and interdependence between energy and environment protection.

In 1982, on June, 5-16, in Stockholm at the UN Conference on Environment Surrounding Human Being the UN member-countries agreed that it was necessary to take urgent measures to solve issues of environment degradation.

In ten years, in 1992, on June, 3-14, at the UN Conference on Environment and Development, which was held in Rio de Janeiro, the UN member-countries agreed that environment protection and social and economic development played a great role in sustainable development. The Principle 4 of the Rio de Janeiro Declaration on Environment and Development says: «To achieve sustainable development, the environment protection must be the integral part of the process of development and it cannot be rated in isolation from that».

And in ten years on September, 2-4, 2002, at the World Summit on Sustainable Development in Johannesburg, South Africa, the first problem, which participant-countries were facing to, was pointed out as following: «We recognize, that poverty extermination, modification of consumption and producing models, as well as protection and rational utilization of natural resources in the interest of social and economic development are the most important aims and basic needs of sustainable development». And further among the problems, the countries of the world community were facing to, it was marked that: «It is doing damage to the global environment so far. Loss of biodiversity and exhaustion of fish resources are in process of continuing, desertification are embracing more and more fertile lands, fatal consequences of climate change are quite obvious, natural disasters are becoming more and more frequent and more destructive, developing countries are becoming more and more vulnerable, and pollution of air, water and sea environment is stripping millions of people of a worthy life».

As it has been underlined above, the level of a material life, and, finally, of mental culture of people are in direct dependence on quantity of the energy, available at their disposal, so far as the quantity of energy, which people can get only from environment resources, will depend on, as far as the environment is favorable and rich in energy sources. Hence, dependence and interrelation between quantity of energy which people can get to satisfy their all growing needs, and well-being of environment, surrounding people, are quite direct.

Therefore people, being concerned about searching for more powerful, more effective energy sources must always remember, how their utilization will be reflected on environment and themselves, because they are part of the environment. They should take care of that that this reflection has positive character, but by no means negative one, causing damage to the environment and therefore to them.

In the environment, surrounding people (environment for living and industrial activity of people), as a rule, the following basic components are pointed out:

- air environment (atmosphere);
- water environment (hydrosphere);
- fauna (human being, domestic and wild animals, including fishes and birds);
- flora (cultural and wild plants, including growing in water);
- ground (vegetative layer);
- bowels (top part of the earth's crust within the limits of which mining operations are possible);
- climatic and acoustic environment.

Not long ago it was considered that the most vulnerable components, without which the human being could not live and which were damaged greatly due to people's activity, connected with industry development and urbanization, were air environment (atmosphere) and hydrosphere. But today, together with development of new technologies and constantly growing demands to satisfy people's needs in energy, considerable damage is causing to all spheres of environment. For example, burnt fuel (fire wood, coal, black oil, gasoline, etc.) evolves huge quantity of persistent organic pollutants into the air environment, which with precipitations and wind are distributed to extensive territories, polluting ground, water, flora and fauna.

How to mitigate the impact on the environment, surrounding people, and whenever possible to make the utilization of energy sources safe – this idea must be in our mind every minute and correct our activity when we aspire to satisfy our needs in energy.

The world history of energy was developed unfortunately in such a way that people, after having exhausted wood stocks, discovered more high-calorie kinds of fuel as coal, oil and natural gas, and staked on them. In other words they staked on fuel reserves in the interior of the Earth. But, on the one hand, the reserves of oil, gas, coal are not at all endless. To compose all these reserves, nature has needed millions of years, but they will be utilized within hundreds of years. At a modern level of energy resources consumption, oil will last 50 years, natural gas – 73 years, coal – 170 years, brown coal – 500 years. Today people in the world have begun to think more seriously about how to prevent injurious plunder of the Earth riches. We know that the reserves of fuel will last for many centuries if only people act in such a way.

But ecologists are uneasy not because of the reserves of fuel, though it is also of great importance, but because of mining and incineration of them are causing colossal damage to the environment. Vast territories quite suitable for agriculture are becoming unfit for usage at a long date. In spite of the fact that methods of fuel incineration have become much more complex and perfect, the increasing demands to environment protection demand new ways of looking at energy.

The Central Asia countries, including Kyrgyzstan, are among those countries, which actively use <u>mineral fuel</u> for power supply. It is paradoxical, but, fortunately, in the Central Asian region there is no much forest, so that it is not substantial recourse of energy. For example, in Kyrgyzstan only about 4 percent of the territory is covered with woods. About the same situation is in other countries of CA. In their arsenal there are coal, oil and natural gas. Hence, the Central Asia region is experiencing great impact on the environment from products of organic fuel incineration. But, being the developing countries, they cannot spend enough money for working out and application of more perfect, less harmful methods of fuel incineration.

Let's give one simple example. In Kyrgyzstan there is no enough high-calorific coal, and the need in it is great. But there are sufficient deposits of low-calorie coal, which produced much ash after incineration.

Furnaces, which are used for incineration, are out-of-date design and do not allow to apply more perfect methods. There are special furnaces designed for low-calorie coal, which allow reduction of emissions in the atmosphere and quantity of ashes. They are being produced at one plant not far from Moscow in Russia. But the Kyrgyz Government has no money to buy them and install instead of the old furnaces in all public establishments of Kyrgyzstan. Therefore they will probably shed harmful substances into atmosphere and pollute land with ashes for a long time.

The second most popular energy source in the Central Asia countries is <u>waterpower</u>. Kyrgyzstan and Tajikistan are among the first ten countries of the world concerning waterpower resources. In these countries the hydro-power engineering, especially big one, has developed greatly in the '70s and in the first part of the '80s the last century. The Government of the Soviet Union (these two Republics formed part of the SU) was concerned about how to supply with energy the region and the whole country, but thought very little about ecological consequences of huge transformations in that region. Thousand and thousand hectares of the fertile lands had been flooded under water basins. The climate in the big areas around the hydro-power stations had changed because new big water reservoirs had been built. Many animals and birds had migrated, the biodiversity had been reduced. Along the whole extent of high-voltage network's lines there was steady electric field which made pernicious impact on the whole alive nature.

In the world up to the midst of the '80s years of the last century the <u>atomic engineering</u> was developing dynamically. And it was considered that it would become the basic alternative to the energy produced by organic fuel incineration. This process actively affected the Central Asia countries, especially Kyrgyzstan. Kyrgyzstan is rich in uranium ores. They began to be extracted and processed into fuel for Soviet atomic power stations, military-industrial complex and space industry's needs, but uranium wastes were stored in tailing dumps on the territory of Kyrgyzstan. Today in different parts of the country there are tens and tens tailing dumps with uranium wastes, which are representing enormous threat to the environment not only in Kyrgyzstan, but also in the whole Fergana valley and far more then the valley. The tailing dumps management, re-vegetation are needed in much sum of money, but the country does not have it.

According to the premises it is obvious that the most popular energy sources, which are the most popular in the Central Asia countries, prove to be not in big friendship with environment protection.

How should the situation be changed? How should constantly growing needs in energy be adjusted with environment protection concern?

How can we minimize the negative impact of producing and utilization of energy for people's needs on the environment?

The progressive mankind has found the answer to these questions in utilization of renewable sources of energy (RSE). They are:

a. solar energy;
b. wind power;
c. energy of the small springs and currents;
d. geothermal energy;
e. biomass (household, agricultural wastes, wastes of live-stock breeding, poultry farming, wood, wood processing industry, etc.);
f. low-potential heat (diffused heat of air, ocean water, seas and reservoirs).

All variety of RSE is produced from three global kinds of sources: sun, heat of the Earth and energy of orbital movement of planets. The sunlight capacity surpasses all others sources of more than 1000 times.

It is known, that practically all kinds of RES but deep geothermal energy and biomass, alongside with obvious advantages, have essential lacks: low specific potential (dispersed) and the irregularity of receipt which depend on climatic conditions, daily and season cycles. Therefore to utilize them effectively it is necessary to solve a number of engineering tasks on creation of economical and reliable equipments and systems which can take in, concentrate and transform these kinds of energy into thermal, mechanical and electric energy adequate to the consumer's needs.

But in this case we are interested not in the technical issues of the question though it is also of great importance, but in **the ecological issues**. The basic advantages of the RES are inexhaustibility and ecological cleanliness. Their utilization does not change power balance of the planet, it considerably reduces pollution of atmospheric air, ground and water basins, essentially softens impact on animate nature. These very qualities have served as the reason of rapid evolution of renewable energy in the developed countries and are drawing increasing attention of developing states.

Let's consider in short ecological advantages of some RES, which are relevant to people living in the Central Asia countries.

Solar energy. The potential opportunities of the power engineering, based on utilization of sunlight, are extremely great. It must be noted that if we use only 0, 0125 % of the quantity of solar energy we can satisfy all modern needs in energy of all people in the world, and if people use only 0,5 % of it they can satisfy their needs in energy for the perspective future.

Unfortunately, we doubt whether we can manage to realize these great potential resources in a big scale. One of the most serious obstacles of such realization is, as it is already marked above, low intensity of sunlight.

The leading countries to use solar energy are Japan, the USA, Switzerland, Germany and Israel.

Today among the main ecological advantages of solar energy utilization there are:

- utilization of solar energy considerably saves other traditional kinds of energy resources (organic fuel), which are used for heating buildings, houses, etc. and supplying with hot water economic needs;
- opportunity of independent power supply, which is especially important for places of incompact residing of the population, for example in mountain areas when the solar energy, transformed into electric energy, does not require in electric line system for long distances not doing damage to landscape and biodiversity;
- solar installations practically do not need operating costs, are out of fix and demand expenses only for construction and maintenance in cleanliness, they can operate extremely indefinitely, hence, it is possible to get economy of other natural resources, which are necessary to produce energy.

The deficiency in utilizing solar energy today, negatively reflecting on environment protection, so far is that that solar energy belongs to the most raw material intensive industry, hence it requires much aluminium, for example. Large-scale utilization of solar energy will lead to great use of raw materials, and, therefore manpower for mining raw materials, enriching, getting necessary materials, constructing heliostats, collectors, other equipment, their transportation.

Wind energy. There is a large amount of moving air. Reserves of wind power are more than a hundred times exceed reserves of water power of all rivers on our planet. Why so plentiful, accessible and ecologically clean source of energy is utilized so little? Today the engines using wind power cover only one thousand global needs in energy.

In the countries, where there are natural steady winds, the target of utilizing wind energy has become different. It is for producing electric power, and it has developed greatly as, for instance, in the Scandinavian countries and some other countries of the continental Europe.

In the Central Asia countries this source of energy is used comparatively very little. For example, in Kyrgyzstan in the places where wind potential is 10-12 m/sec, there is no consumer. It is in high-mountainous passes, canyons and very remote places. Utilization of wind energy in such conditions is economically and socially unprofitable. But where there is a consumer, the speed of wind does not exceed 3-5 m/sec. To use it, is also unprofitable from economic point of view.

The main ecological advantage of wind energy is that it is a non-polluting energy source, which is transformed into either mechanical or electric power. In some cases its mechanical power can be used to save electric power. For example, we can utilize a simple wind wheel with blades to get water from rivers and other relevant water reservoir for irrigation instead of using electric pump.

But there are some ecological interdictions in utilizing energy of wind. Wind installations, especially big ones and if they are many in one and the same place, make significant noise. The acoustic environment, as it has been marked above, is one of the components of environment, surrounding the people. Excessive noise is the considerable factor, which negatively affects human people and animals' health. The experiments have proved the fact that increased noise adversely affects even the plants. It is known, that in radius up to ten kilometers from wind turbines biodiversity gradually disappeared. And people felt bad living near to wind turbines. Noise and vibration of air disturbed people when they worked, rested, slept.

Scientists are working hard on designing new wind turbines which can considerably reduce noise but how to get rid of vibration of air streams – that is a problem.

Energy of small rivers and water-currents. For mountain areas power of small rivers and water-currents are the most accessible and effective energy sources. Taking into account that mountains occupy one fifth of the land, one can assert that it is potentially accessible and effective energy source for great number of people living on the Earth and that means for those who do not utilize much energy resource because they live far from centralized systems of power supply. They need energy mainly for household demands and manufacture. They are shepherds, geologists, beekeepers, farmers, workers of forest reserves, etc.

Ecological cleanliness of this energy source is blameless. Practically there is no ecological interdiction for its utilization. Utilization of small water-currents energy transformed into electric power with capacity 1, 5, 16, 22 kilowatts allows to reduce cutting down of woods and utilization of wood as fuel, to mechanize significant part of economic works at farms, to develop family life culture in the remote mountain areas.

In Kyrgyzstan, for example, small-lot production of micro-hydroelectric power stations with the capacity of 1, 5, 16 kw has been arranged. A micro-hydroelectric power station is a small portable flexible-hose equipment for energy supply of seasonal consumers and cattle breeders.

Stationary micro-hydroelectric power stations and small hydroelectric power stations are also utilized.

Biomass energy. Special ecological value is represented with utilization of energy of biomass. One can dedicate odes to biomass as energy resource. Biomass is organic wastes of animal industry, agriculture, household, wood industry, etc.

Marsh-gas extracted from biomass according to well known technologies is an excellent, non-polluting fuel, which is used:

- first, to get thermal energy that is used for heating premises, houses, preparing food, heating hot water, etc;
- second, to transform and produce electric power;
- thirdly, to use as fuel for cars and other automobiles that allows significantly to reduce persistent organic pollutions into the atmospheric air in comparison with utilization of oil products.

But the utilization of biomass for producing energy has not only direct ecological power effect. This is only one advantage of biomass ecological value.

People can get highly effective organic, non-polluting fertilizer – methanoic fluent. It is the result of cattle-breeding, poultry-farming and other similar agriculture waste products and effluents of city water drains fermentation and gathering marsh-gas. The fertilizer is used according to very simple technology - together with irrigation. It essentially raises productivity of agricultural crops, improves land structure. Hence, we can reduce fuel utilization by tractors, machines, etc. to process fields and applying other fertilizers.

With reference to the Central Asia countries biomass can be used as an energy source almost everywhere, both in cities and in countryside, including the remote mountain areas. For example, in Kyrgyzstan, there are several experimental biogas reactors: small ones for two-three houses, a bit bigger ones for one street in a village and one comparatively big one at a pig farm. They are successfully approved. Information and educational activity were done among villagers. This initiative has very good prospects. Some projects on using biomass as energy source are under consideration.

As for the **geothermal energy** it should be noted that there are many places in the Central Asia countries with geothermal resources. The geothermal energy mainly is used in the form of geothermal water. In many places with geothermal water health resorts and sanatoriums were built. These health resorts and sanatoriums, as well as local people who live there, are using geothermal water as medical treatment to restore health. But it is also used to heat premises, houses, as hot water for household needs.

They say that it is one of the wonderful presents of the God.

Concerning utilization of **low-potential heat** (diffused heat of air, ocean water, seas and reservoirs) it is obvious that for developing countries of Central Asia it will be rather distant future. As for the scientific investigations this theme can be developed, but as for practical realization it is a dream so far.

In conclusion it should be underlined that there are very optimistic prognoses concerning the utilization of renewable energy resources in the Central Asia as well as in the other parts of the world. It is promoting the solving of three global problems mankind is facing today – energy safety, environment protection and food safety.

Chapter 16

TO THE PROBLEM OF PRODUCTION AND USING OF BIOFUEL IN CONDITIONS OF UZBEKISTAN

O.V. Lebedev, R.K. Musurmanov, K.A. Sharipov, A.S. Azizov

Academy of Sciences of the Republic of Uzbekistan

For last hundred years not times were expressed opinion on production opportunity of fuel on the basis of reprocessing biological materials. During the years repeatedly came back to this problem. Various periods experts give their conclusion on exhaustion of natural stocks of liquid hydrocarbons. Within the last several decades use of ethanol and vegetable oils was propagandized as fuel for engines of internal combustion. However, attempts of production of stocks of soil that reduced the value of such experiments.

It is obviously that, on account of energy and fuel, produced from biomass it is impossible to satisfy completely energy needs of the advanced industrial countries, however even that small share of energy (about 6-15%) which can be covered on accound of biomass is worthy for attention. Various kinds of biomass in all regions are practically available, and almost in each of them it is possible to produce energy and fuel from biomass.

It is solved energy and ecological problems at application biofuel on the basis of bioethanol as motor fuel.

Ethanol possesses a number of positive qualities: it is not toxic, manufactured its production, some qualitative characteristics of ethanol are higher than other kinds of liquid motor fuel.

It is pointed by foreign scientists, who engaged in fuel saturated with oxygen, spirit can mix up or turn into emulsion (also with diesel fuel) with high quality of ignition. Pure ethanol completely mixs up

A. Iacomelli (ed.), Renewable Energies for Central Asia Countries: Economic, Environmental and Social Impacts, 175–182.

with diesel fuel at the certain temperature. In conditions of low temperature or when the water content exceeds the determined level by us miscibility is limited.

Methanol, even when it is dry, not completely mixs up with diesel fuel and can be corrosive.

Global manufacture of ethanol as fuel was doubled in 1996 and can be doubled again by 2010. For last years consumption of has essentially increased, but in comparing with oil it still remains low. The greatest manufacture and use of ethanol belongs to the USA and Brazil – where volumes of use are many times higher than in any other country. But even in the USA ethanol represents less than 2% of transport fuel (while in Brazil ethanol makes approximately 30% of for gasoline).

The increase in use of biofuel solitarily influences ecology and hotbed effect, reduces issue of.

As fuel ethanol produces from grain (in the USA and Europe), waste products of sugar production (Europe), from sugarcane, from biomass of waste products of animal industry and fruit-and-vegetable growing (the countries of the European Community).

Biofuel is less poisons than usual fuel of an oil origin, but it can sometimes result in issue of aldehyde from use of ethanol. Ethanol can be used as additive to increas octan number of gasoline. In some countries of Europe before to be mixed with gasoline is transformed to tertiary bulyl-ethyl ether.

Biofuel can be easier in the commercialization process than other alternative fuel. Therefore its use increases in today's vehicles, with the purpose of decreasing of global consumption of oil by 10%. For this purpose, for example. In the USA the subsidy is entered at a rate of 0,14$ per liter to support production and sale of ethanol obtained from grain. Also the marked of biofuel in Canada is supported.

Biofuel represents fuel ethanol, biogas or hydrogen, which produces from consecutive circuit of production, transportation to the pumfying plant, transformation into final fuel and transportation to the filling station.

In 2004 in the USA produced more than 10 billion litres (2,6 billion gallons) fuel ethanol. All gasoline sold in Brazil, contains 22-26% of ethanol.

Development or world production of ethanol for the period 1975-2004 is shown in the following figure:

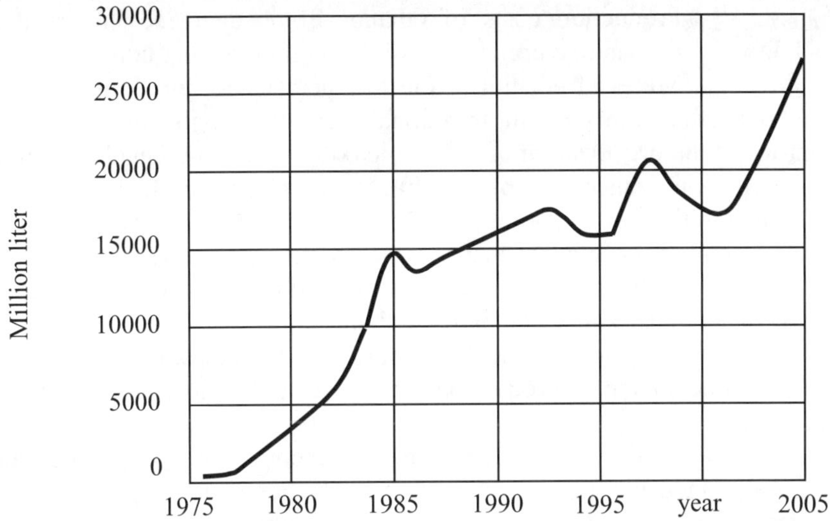

Figure 1. World production of ethanol.

High solar radiation conditions in Uzbekistan and good productivity promotes to produce high crops of fruit-and-vegetables at low expenses. Therefore, from the waste products of these products (tomatoes, apples, grapes, etc) it is possible to produce cheap fuel ethanol.

Ethanol can be formed of any biological raw material for the industry which contains considerable amounts of sugar or materials capable to be transformed to sugar, such as starch or cellulose.

The fermentation is transformed sugar with six atoms of carbon (mainly glucose) into ethanol. With this purpose, millions dollars are spending for researches on improvement et fermentation processes of hydrolysis. From a ton of sugarcane waste it is possible to produce 360-470 litre of ethanol and 346-385 litres from grain. (Wang, Levy, Marland, Levington).

The American Energy Ministry has more than 100 million dollars from the budget under program FY 2003 on processing biomass into fuel.

In Canada scientific researches on ethanol are successfully carried out according to the program of technologies of renewed energy sources. (RETP)

One litre of ethanol equivalent to gasoline is equal on energy of oil of 0,85-0,88 litres: Achievement in the field of production and use of biofuel in the world are coordinated by the international Energy Association

(IEA), including the USA, Canada, twenty European countries, Japan, South Korea, Australia and New Zealand.

Production costs of ethanol from sugarcane in Brazil are much lower than manufacture of ethanol from grain in other countries.

Mixes of ethanol of the low percentage, for example, marks E-10 (10% ethanol mixed with traditional fuel) are available on sale in many countries of the organization of economic cooperation and development. So, 45% of fuel in France, 7% of fuel in Germany, 20% of fuel in the USA represent additives of ethanol in diesel fuel or gasoline (biofuel). This fuel is more favorable for a climate than oil fuel, have lower issue CO2 and other gases creating hotbed effect. Biofuel can be used in system of a diesel engine as 5%, 10% or 20% a mix of ethanol with diesel fuel.

In agriculture of the Republic of Uzbekistan it is possible to count a prime source of production of fuel ethanol from waste products of food manufacture.

Being large branch of production of goods, the food-processing industry processes a significant amount of grapes, fruits, vegetables, sugar beet, grain and other agricultural raw material which predetermines the further development of reprocessing branch.

By preliminary calculations at processing agricultural products on canning and wineries in Republic is accumulated more than 300-350 thousand tons of waste products. By additional reprocessing these waste products, in the incorporated devices, offered by authors, it is possible to produce 20-25 thousand tons biofuel or additives to traditional fuel can be solution of power and ecological problems of the Republic.

The basis stage in production of bioethanol is the process of fermentation, in which, productivity of bioethanol completely depends on this process.

The process of fermentation depends on following factors: temperature, concentration of carbohydrates, time, pH of media, existence of oxygen, complex of yeast bacteria. Controlling the pH of the media plays an important role in production of bioethanol.

For decrease pH of biomass, which contains carbohydrates it is offered to add the waste of tomato reprocessing production.

Produced raw bioethanol is dehydrated with the help of new technology-evaporation through membranes, mixtures, such as, high spirits (fusel oils), methyl spirit and complex ethers remain in structure of bioethanol. There fore, bioethanol received such way cannot be used for the food purposes.

The analysis of spirit was carried out for analysing of micropurity and results are given in the table 1: acetealdehyde, ethers, methyl spirit and sum of high spirits in correspondence with GOST-5962 “Ethyl spirit food, rectified with extra purifying”, “Extra”, and “Lux”.

Table 1. Micropurity content of bioethanol

Compound	A unit of measurements	Content
Acetealdehyde	*mg/l*	373,6
Total ethers	mg/l	66,3
Methanol	% total	0,22
Total high spirits	mg/l	396,9

Solutions of diesel fuel with bioethanol (ethyl spirit) are investigated in various concentrations: 1%, 2%, 3%, 4%, 5%, 6%, 7%, 8%, 9%, 10%.

Researches are carried out in laboratories of FCI I «UZLITINEFTGAZ».

Diesel fuel with bioethanol forms emulsion, which with rising of temperature is stratified, and then forms true solutions. The output of water in a mix changed depending on concentration of bio ethanol in diesel fuel: from traces up to the 1%.

Results of research are showed in a graphical form:

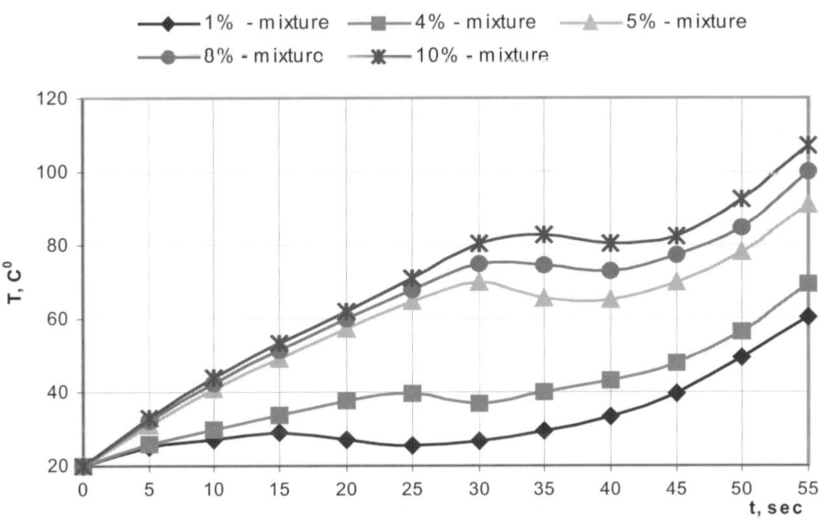

Figure 2. Dependence of temperature of miscibility on line.

Results of research of these mixes in a graphic form are submitted in figure 3. As shown in figure 3 density and viscosity of solutions of mixes decreases with increase in temperature.

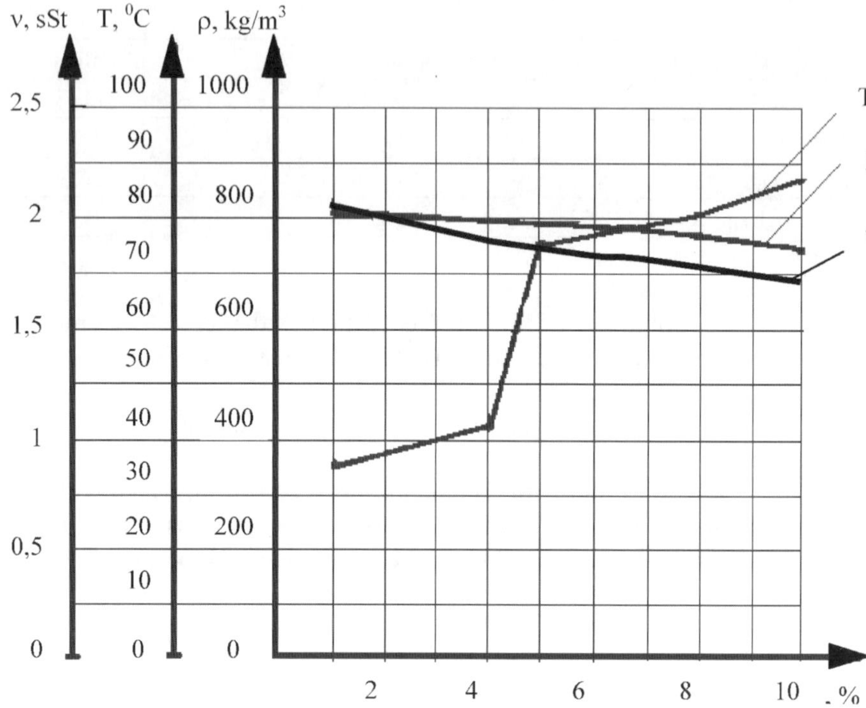

Figure 3. Dependences of temperature of miscibility T, density ρ and viscosity υ from various concentrations of bioethanol.

However, it is necessary to remember that increasing of amount of bioethanol in diesel fuel results in increase of water content (on GOST requirement presence of water is in admissible) as it renders corrosive attacks on metal.

In connection with arisen problems, speed of corrosion of metal with 4% and 8% solutions of spirit in diesel fuel investigated, tests for speed of corrosion of diesel fuel for steel CT-20 also carried out.

As researches showed, speed of corrosion of steel in pure diesel fuel was equal to 0,070 g/m2·hour, with 4% solutions of spirit in diesel fuel – 0,084 g/m2·hour, with 8% solution – 0,091 g/m2·hour. Speed of corrosia in diesel fuel was equal to 0,064 mm/year, in mixture – 0,079 mm/year.

Table 2. the comparative characteristic standard of diesel fuel with a mix "diesel fuel : bioethanol"

Parameters	On STATE THAT	Diesel fuel (analyzed)	Diesel fuel: Bioethanol of 8 %
1. The maintenance{contents} of water, %	Absent	Absent	Traces
2. The density, kg / m ³ at 20 °C	Is not normalized	811	810
3. Viscosity, cCт at 20, °C	1,8 – 3,2	2,9	3,0
4. Temperature of flash in open cup	-	64	53
5. Acidity of mg KOH on 100 m л fuel	No more 5	2,1	0, 403
The fractional structure is overtaken at T°C Not above 50% 96% 95%	 250 340	 250 340	 255 340
7. Speed of corrosion Steel CT of-20 г/mm ² * hour	-	0,07	0,091

Tests of diesel engine Д-21 for fuel mix "diesel fuel: bioethanol" were carried out according to the method which we are described. From operational parameters of diesel engine Д-21 in figure 4 change of capacity from concentration of bioethanol in mixture of "diesel fuel : bioethanol".

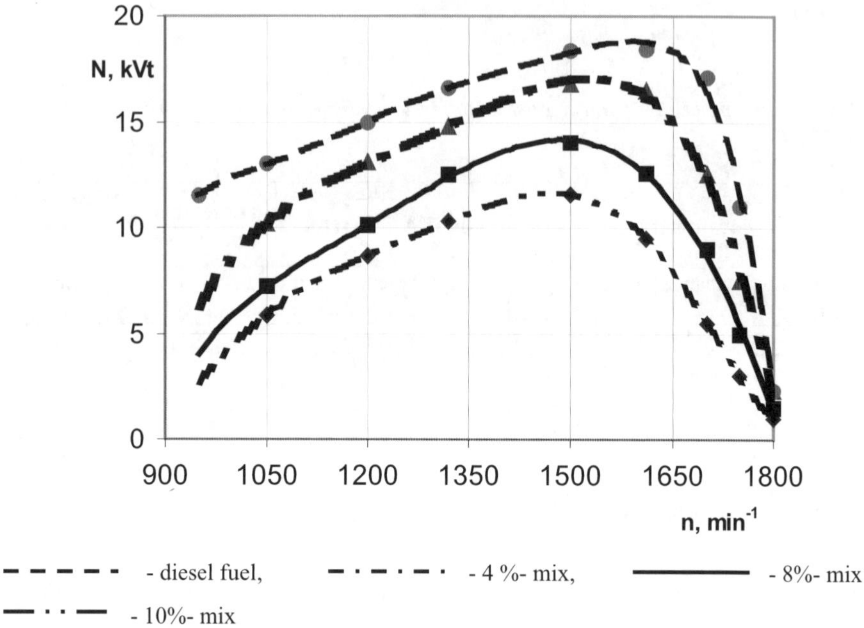

Figure 4. Dependences of power from revolving frequency at various mixtures of bioethanol.

As shown in the schedule with increasing of concentration of bioethanol capacity decreases. But, bioethanol has very high latent heat of evaporation (1100 kDj/kg) in comparison with diesel fuel (210 kDj/kg).

With a view of wide use of biofuel it is necessary to stimulate manufactures in Uzbekistan giving them state grant.

Besides, it is necessary to begin sale of automobiles designed exclusively on biofuel and prices for them must artificially reduce